Bacterial Activation of Type I Interferons

AF587595

Dane Parker
Editor

Bacterial Activation of Type I Interferons

Editor
Dane Parker
Columbia University
New York, NY, USA

ISBN 978-3-319-38502-0 ISBN 978-3-319-09498-4 (eBook)
DOI 10.1007/978-3-319-09498-4
Springer Cham Heidelberg New York Dordrecht London

© Springer International Publishing Switzerland 2014

Softcover reprint of the hardcover 1st edition 2014

This work is subject to copyright. All rights are reserved by the Publisher, whether the whole or part of the material is concerned, specifically the rights of translation, reprinting, reuse of illustrations, recitation, broadcasting, reproduction on microfilms or in any other physical way, and transmission or information storage and retrieval, electronic adaptation, computer software, or by similar or dissimilar methodology now known or hereafter developed. Exempted from this legal reservation are brief excerpts in connection with reviews or scholarly analysis or material supplied specifically for the purpose of being entered and executed on a computer system, for exclusive use by the purchaser of the work. Duplication of this publication or parts thereof is permitted only under the provisions of the Copyright Law of the Publisher's location, in its current version, and permission for use must always be obtained from Springer. Permissions for use may be obtained through RightsLink at the Copyright Clearance Center. Violations are liable to prosecution under the respective Copyright Law.

The use of general descriptive names, registered names, trademarks, service marks, etc. in this publication does not imply, even in the absence of a specific statement, that such names are exempt from the relevant protective laws and regulations and therefore free for general use.

While the advice and information in this book are believed to be true and accurate at the date of publication, neither the authors nor the editors nor the publisher can accept any legal responsibility for any errors or omissions that may be made. The publisher makes no warranty, express or implied, with respect to the material contained herein.

Printed on acid-free paper

Springer is part of Springer Science+Business Media (www.springer.com)

Preface

Type I interferon (IFN) signaling has long been recognized as a critical component of innate immune defense to viral pathogens. It is now established that bacteria too are able to activate this pathway. Typically bacteria activate type I IFN signaling through TLR-dependent mechanisms, through recognition of LPS in Gram negative organisms or via TLRs and cytosolic receptors that respond to nucleic acids and messaging molecules that are either endocytosed or secreted directly into the eukaryotic cells. The consequences of type I IFN signaling on host outcome can be either protective or damaging, depending on the organism.

The signaling behind the type I interferons is a rapidly progressing field. Initial activation of type I IFNs was limited to the innate receptors that had been identified such as LPS via TLR4 or nucleic acids through TLR3 and TLR9. Recent work has uncovered several cytosolic receptors that are able to recognize different microbial products, such as nucleic acids, cell wall components, and signaling molecules. In many cases these different receptors ultimately converge on shared adapter proteins, kinases, or transcription factors that lead to the production of type I IFNs. The activation of type I IFNs through their cognate receptors and JAK/STAT signaling leads to the production of hundreds of gene products that are likely unique to each pathogen.

The activation of type I IFNs differs between each microbial pathogen and also within some species. In many cases activation of the pathway is a passive process, whereby cell wall components such as LPS or peptidoglycan are exposed to receptors or nucleic acids are sensed during cellular destruction. Bacteria also activate type I IFNs by actively assaulting host cells. In these instances toxins or bacterial secretion machinery is designed to lyse or inject effector proteins that alter host machinery allowing microbial patterns to be sensed.

The consequence of this activation also varies by organism. In *Legionella pneumophila* type I IFNs play a positive role, restricting replication of bacteria inside macrophages. This is also the case for some *Chlamydia* species; however, in vivo results have shown the opposite result with reduced bacterial burden in mice unable to respond to type I IFNs. Type I IFNs are detrimental in the context of chronic *M. tuberculosis* infection, while type I IFNs appear to play a negative role in response to several bacteria including streptococci, *Salmonella*, and *Staphylococcus aureus*. In the case of *S. aureus*, different strains signal the production of type I IFNs through different receptors. The activation of similar and divergent pathways within the same species as well as other different species of bacteria makes for interesting comparisons. This book provides an overview of how type I IFNs are activated and the role they play in several important bacterial pathogens, highlighting how the immune response can influence the outcome to infection.

New York, NY Dane Parker

Contents

Production and Action of Type I Interferons in Host Defense

Paul J. Hertzog

Introduction

The interferons (IFNs) are a family of cytokines that function in the host response to environmental stress [1]. The evolution of the IFN response has adapted to perform a wide range of physiological and pathological functions. The IFNs are classified into three types distinguished by amino acid similarity; cognate receptors, through which they signal; and to a lesser extent, the production stimulus and cell. Type I IFNs are a multi-gene family composed of 13 IFNα subtypes, a single IFNβ, IFNε and IFNω, and other species-specific members, produced by most cell types and acting via IFNAR 1 and 2 receptors [2]. Type II IFN has a single member, IFNγ, produced mainly by activated NK and T cells and signaling via IFNGR1 and 2 receptors [3]. Type III or IFNλ has two members, produced by many cell types stimulated by pathogens and acting via IFNL1 and IL10Rβ receptors [4]. This review will focus on type I IFNs, setting the scene for their role in host defence against bacterial infections. IFNs have multiple effects on cells, which include rendering them resistant to viral infection, modulating proliferation, differentiation, survival and migration, as well as other specialized functions [5]. Thus, IFNs can regulate the development and activation of most effector cells of the innate and adaptive immune response. Type I IFNs signal via the JAK/STAT signaling pathway to regulate the expression of genes that encode the effector proteins of the response including antiviral and antibacterial effectors. Their broad effects on a range of target cells, necessitates a fine balance in the IFN response to ensure protection of the host against insult and a return to homeostasis, but avoid potential toxicity or chronic disease. Excessive IFN production contributes to acute septic shock in animal

P.J. Hertzog (✉)
Centre for Innate Immunity and Infectious Diseases, MIMR-PHI Institute,
Monash University, Clayton, VIC, Australia
e-mail: paul.hertzog@monash.edu

© Springer International Publishing Switzerland 2014
D. Parker (ed.), *Bacterial Activation of Type I Interferons*,
DOI 10.1007/978-3-319-09498-4_1

models, and long-term deregulation of type I IFN signaling contributes to the pathogenesis of autoimmune diseases such as systemic lupus erythematosus. Understanding the regulation of type I IFN production and the actions of this family of proteins on cells is necessary to gain insights into their role in the pathogenesis of bacterial infections. In some cases, particularly with extracellular pathogens, IFNs are protective, whereas they increase susceptibility to intracellular pathogens.

Production of Type I IFNs

The production of type I IFNs was first described in response to viral infection and remains best characterized in response to these pathogens [6–8]. Nevertheless, it is increasingly evident that type I IFN production is activated by a wide variety of stimuli, including bacteria [9], physiological stimuli [10, 11] and cancer cells [12, 13]. The deluge in information characterizing the pattern recognition receptors (PRRs) that sense "danger" signals has provided considerable explanation of the mechanism whereby type I IFNs are produced [14–16]. Once PRRs bind ligand, they engage intracellular signaling molecules, often specific for the PRR family involved, and then activate kinases that in turn activate a restricted range of transcription factors such as NFκB and the interferon regulatory factors (IRFs) that stimulate the induction of pro-inflammatory cytokines and type I IFNs, respectively. The IRFs are a nine-member family of latent transcription factors involved in type I IFN production (IRF1, 3, 5, 7) and signaling (IRF9), among other functions [17]. As discussed in detail below, IRF3 is activated by many PRRs to induce IFNβ gene expression (in conjunction with NFκB and AP1) but not the expression of IFNαs. On the other hand, IRF7 is also activated by many PRRs, but can activate expression of IFNβ and IFNα subtypes. In addition, IRF5 and IRF1 appear more restricted in their upstream activation pathways and these also activate IFNα gene expression.

TLRs 1, 2, 4, and 6 are cell surface PRRs that sense cell surface or secreted ligands, or pathogen-associated molecular patterns (PAMPs). The TLR4 signaling pathway activated by Gram negative bacterial lipopolysaccharide (LPS) in complex with MD2 and LBP is the best characterized and prototypic PRR signaling pathway. Ligand activated TLR4 engages four TIR domain-containing adaptor molecules: MyD88 and Mal, which activate the NFκB pathway, and TRAM and TRIF, which activate the IRF3 previously phosphorylated upstream by TBK and IKKε [14–16]. This pathway, in conjunction with NFκB, activates expression of IFNβ specifically, since this is the only type I IFN with neighboring IRF3 and NFκB binding elements in its promoter. *Escherichia coli* and *Salmonella* are strong activators of TLR4, whereas other bacteria such as *Helicobacter pylori* produce LPS that only weakly stimulates TLR4, which may explain their relative virulence [9].

TLR2 usually acts as a heterodimer with TLR1 or TLR6 and recognizes different peptidoglycans to activate the NFκB pathway driven pro-inflammatory cytokines via MyD88 and Mal. This signaling pathway is not usually associated with activation of IRFs and IFN production. However, exceptions have been reported [18], including a study involving the commensal *Lactobacillus* [19], but the details of the pathways remain to be fully elucidated.

TLRs 3, 7, 8, and 9 are endosomal sensors of nucleic acids including dsRNA (TLR3), ssRNA (TLR7/8) and bacterial CpG DNA (TLR9). TLR3 is the only family member that does not utilize MyD88, but signals via TRIF. These endosomal TLRs recruit adaptors and activate TBK and/or IKKε, which in turn activate IRF7 and 3 to drive the induction of IFNαs and IFNβ [16]. TLR9 senses *Staphylococcus aureus* and activates IFN production via IRF1 [20]. Group A and B *streptococci* are recognized by TLR7 and activate IFNs via IRF1 [21].

The RIG I-like family of receptors (**RLRs**) including RIG-I, MDA5 and LGP2 were originally identified as cytosolic sensors of viral 5′-triphosphorylated RNA [22]. Once activated, they are recruited to mitochondria or associated membranes, bind adaptors MAVS/IPS and subsequently activate TBK/IKKε, which phosphorylate IRFs, which themselves translocate to the nucleus and induce expression of IFNα and IFNβ genes [23, 24].

STING was discovered as a molecule that mediated the induction of IFNβ in response to cytosolic DNA from pathogens or necrotic cells [25]. Subsequent studies cast doubt on whether the endoplasmic reticulum-localized STING directly bound DNA (reviewed in [26]). It was found that STING was the receptor for cyclic di-nucleotides such as c-di AMP or c-di GMP which act as PAMPs, for example in macrophages infected with *Listeria monocytogenes*, after listerolysin O-mediated (LLO) their release from vacuoles, possibly via DDX41 [27–29]. Another STING activating PAMP is cGMP-AMP generated by the IFN-inducible enzyme cGAS, which is important in sensing cytosolic DNA and initiating the innate immune response to pathogens. DNA from *Chlamydia muridarum* [30], *Myocbacterium tuberculosis* [31] and *Legionella pneumophilia* [32] have also been shown to activate STING and induce IFNβ expression.

Cytosolic Sensors

DAI was the first reported cytosolic sensor of DNA from viruses or bacteria, inducing IFN via TBK and IRF3 [33]. DAI senses *Streptococcus pneumoniae* [34]. Another study showed that **DNA-dependent RNA polymerase III** converts cytosolic DNA into RNAs that act as PAMPs to activate RIG-I [35]. DNA released into the cytosol during infection with *Francisella tularensis* is sensed by the **AIM2** inflammasome which in turn activates IRF3 and type I IFN production [36, 37]. *L. monocytogenes* also activates the AIM2 inflammasome [38]. **NOD 1 and 2** have been speculated to induce IFN production in response to sensing muramyl dipeptide (MDP) from organisms including *M. tuberculosis* [39–41].

Thus, the various PRRs constitute a repertoire of sensors, strategically located through evolution, at different subcellular locations to ensure the detection of a pathogen component, be it outside the cell, in endosomes, free in the cytoplasm, associated with organelles or in the nucleus. The various PRR signal transduction pathways activate one of the IRFs (1, 3, 5, or 7) and occasionally NFκB, to bind promoter elements in type I IFN genes. To complement the upstream signaling pathways, the promoters of the 13 IFNα subtypes and IFNβ genes each contain a distinct number and arrangement of transcription factor binding sites to ensure that one or some of these essential cytokines are produced in response to infection with a broad range of pathogens—both viral and bacterial. This promoter diversity is also likely to be important in determining the IFN subtypes produced by different cell types. A thorough investigation of the many transcription factor binding sites in the promoters of the various type I IFN genes is yet to be performed. However, type I IFNs are not only produced by haemopoietic cells as traditionally thought (originally called "leukocyte" IFN), and recent studies have brought attention to their production by epithelial cells as well [42, 43]. Depending on the expression of signaling molecules, different cell types will differ in their pathways, the repertoire of type I IFNs and the amounts they produce. For example, plasmacytoid dendritic cells (DC) express high levels of constitutive IRF7 and therefore rapidly produce high levels of IFNα compared to other cells. Other cell types respond slower because signaling molecules like IRF7 have to be first induced by IFN.

Two decades of studies in mice deficient in Ifnar 1, through which all type I IFNs signal, have demonstrated the crucial role of this family of cytokines in sculpting the response to viral and bacterial infections [44, 45]. Consistent with this scenario, type I IFNs are never all produced, rarely singly (except IFNβ discussed below) and usually in subsets: for example, some IFNαs +/− IFNβ.

In addition to the mammalian cell components, properties of the pathogen also determine the nature of the type I IFN response. For viruses, whether they constitutively harbor RNA or DNA, single or double stranded, determines the cellular PRR response. Pathogens also activate different PRRs depending on their cellular niche. For bacteria, whether they are intracellular or extracellular pathogens and the nature of virulence factors (such as pore-forming toxins) that might be necessary to "release" PAMPs to the responding cellular compartment, will determine the nature of the response.

Type I IFN Signaling:Receptors

All type I IFNs characterized to date transduce signals via interaction with the receptor components, IFNAR1 and IFNAR2. IFNAR2 is the high affinity binding chain and can be differentially spliced to produce a "long" form which transduces signals (IFNAR2c); a truncated transmembrane isoform that contain little or no signaling capacity (IFNAR2b); and a soluble form (IFNAR2a). IFNAR2a and c isoforms are conserved between human and mouse, whereas IFN2b is specific to humans [46–48]. Studies in the murine model have demonstrated that *in vitro*, soluble IFNAR2a has the capacity to either block signaling or facilitate signaling via a process called trans-signaling whereby soluble receptor binds ligand and presents it to the signaling receptor chain [49]. This process can be a major form of signaling as for IL6, but remains to be determined for the type I IFNs. is [50]. In vivo studies have recently indicated that soluble IFNAR2a does not block responses to IFNβ [51]. The IFNAR 1 chain has been shown to have very low affinity for binding type I IFNs (with one exception, discussed below), but combines with IFNAR2 to generate a high affinity trimeric complex. IFNAR1 is essential for transducing signals for all type I IFNs characterized so far, as determined from numerous studies of IFNAR1 deficient mice. Both receptors appear to be expressed broadly making most cells responsive to IFN, but there have not been extensive studies on the surface expression of receptor components on individual cell types or at different stages of the host responses.

Type I IFN Signaling: Signal Transduction Pathways

Once ligand engages the receptors, the IFNAR1-associated TYK2 and the IFNAR2-associated JAK1 kinases are activated and phosphorylate receptor tyrosine residues [52, 53]. These form docking sites for signal transducers and activators of transcription (STATs), which are themselves phosphorylated, dissociate from the receptors, dimerize and translocate to the nucleus via interaction with importins, and activate the transcription of IFN-regulated genes (IRGs) [54]. Studies have shown that the docking sites for STATs are in the IFNAR2 component of the receptor [52, 55, 56]. The canonical transcription factors of the type I IFN pathway is ISGF3 (composed of a STAT1:STAT2) and IRF9 (also called p48 or ISGF3γ because it is induced by IFNγ). Nevertheless, type I IFNs also activate STAT3, and dimers of STAT3:STAT3 or STAT1:STAT3 can bind GAS sites (interferon-gamma activated sites) in IRGs (sometimes wrongly thought to be IFNγ-specific) [54]. Indeed, type I IFNs can activate all STATs (4, 5, and 6) depending on the cell type. Indeed in PBMCs, STAT5 is the main STAT activated [57, 58].

There are other signaling pathways activated by type I IFNs. Indeed JAK kinases have other substrates [59] and also function to stabilize the IFNAR1 at the plasma membrane [60]. STAT independent signaling was reported for both type I and type II IFNs in STAT-deficient cells using transcriptional profiling [61], but the signaling pathways responsible were not pursued in those studies. Numerous “alternative” type I IFN signaling pathways have been described, including MAPK (p38 and ERK), NFκB and PI3K/AKT pathways [57, 62]. The best characterized of these is the p38 and Erk MAP kinase (MAPK) pathways, which modulate IRG mRNA translation via activation of Mnk kinases [63]. Activation of AKT/mTOR (mammalian target of rapamycin) signaling is also initiated by IFNs, impacting on translation of IRG mRNA [64, 65]. The relative contribution of these and other alternative IFN signaling pathways is likely to be cell and context dependent. For example, type I IFN signaling in T cells has been reported to utilize T cell receptor signaling molecules for antiproliferative activities [66].

Type I IFN Signaling: Interferon Regulated Genes

There have been many studies documenting the nature of IRGs individually for past decades and more recently by transcriptional profiling by microarrays. In an attempt to capture an overview of the response, we have catalogued available microarray datasets of IFN treated cells or organisms; the data is reanalyzed and annotated then placed in a searchable database called the Interferome (v 2.0 http://interferome.its.monash.edu.au) [67, 68]. This represents a tool for identifying a gene as an IRG, or more importantly, for searching a dataset for IRGs. This collection has identified more than 2,000 IRGs (more depending on the statistical cut-off applied) across species, IFNs, and tissue types. These genes encode the effector proteins that mediate the different biological activities of the IFNs. The number of genes in any given condition is usually smaller, often hundreds, and there is considerable difference between different cells or tissues. There are overlaps between type I, II, and III regulated genes and some apparently IFN type-specific genes, although comparisons are often difficult because of differences in experimental conditions [69]. Tools such as the Interferome are important in finding IRG "signatures" associated with disease that might represent modulation of a particular pathway. We have used Interferome and associated tools to identify an IFN signature activated in HIV infected dendritic cells (a gene set regulated by IRF1, despite HIV suppression of IFN production [70]) and another gene signature suppressed in breast cancer metastases (regulated by IRF7; [13]). Interestingly, an IFN signature has been characterized in latent *M. tuberculosis* infection that appears to correlate with disease pathogenesis and is consistent with studies in animal models showing a role for type I IFN signaling in susceptibility to this pathogen (reviewed in [71]).

The best characterized IRGs are those involved in protecting cells from viral infection; individual ones such as 2′–5′ oligoadenylate synthetase, PKR and Mx proteins, having been well characterized for many years [72]. Recently, elegant, comprehensive screening studies of 350 IFN inducible genes have highlighted new IRGs with direct antiviral activities [72]. The studies may inform similar rationales for characterizing the effector functions (anti-bacterial, immunoregulatory) of the many other IRGs. In broad terms the broad repertoire of antiviral IRGs has the ability to restrict different stages of the viral life cycle and different types of viruses, providing broad protection against infection.

Type I Interferon Regulated Antibacterial Responses

Unlike the antiviral effects of type I IFNs, the effects of IFN on bacterial infections are relatively poorly characterized. In general, type I IFNs are protective against extracellular bacterial infections, yet exacerbate infections with intracellular bacteria. This is at least in part due to the differences in organ and cell specificity, the direct effects of IFNs, the impact on cell survival and indirect actions via regulation of the innate and adaptive immune responses [43, 73, 74]. Examples of direct acting antibacterial IRGs include iNOS, NADPH oxidase, nox-2 [73, 75]; TRAIL [76]; and phospholipidscramblase 1 (PLSCR1) [77]; GTPases [78].

Type I Interferon Regulated Immune Responses

There are many different IRGs, intrinsic or extrinsic to immune cells that can affect the trafficking, development, differentiation, survival, and activity of most innate and adaptive immune cells in response to infections, cancer, and inflammatory diseases (reviewed in [74, 79]). Particular cells and responses have been documented to be important in the response to bacterial infections. TNFα and IFNγ up-regulation by type I IFNs increases protection from *S. pneumoniae* infection [21]. Repression of type I IFN induced chemokines CXCL10 and CCL5 reduces cells neutrophil infiltration and impairs clearance of *Pseudomonas aeruginosa* from infected lungs [80, 81]. By contrast, type I IFNs suppress the production of other chemokines such as CCL2, CXCL4 and CXCL9, which recruit monocyte/macrophage and neutrophils [75] leading to exacerbation of infection by *C. muridarum.* Other IRGs include cytokines that activate or repress immune responses including IL10 [82], IL27, and IL17 [83] and FOXP3 which is important in Treg function [74]. Another IFN induced effect that is important in regulating responses to bacterial infection is the induction of apoptosis in infiltrating cells [84]. It is well known that IFNs can regulate the expression of different cell death pathways including bcl-2 and bcl-X [85] and caspase 11 [86] and that IFNs play a role in mediating necroptosis of *Salmonella typhimurium* infected macrophages [87].

Cross-Talk, Feedback, and Feed Forward

There have been numerous publications about cross-talk of type I IFNs with other systems. In general terms, many of the receptors and signaling components of other signaling systems are, in fact, IRGs and the positive or negative regulation of these factors underlie the basis of cross-talk [88]. These include other cytokines (reviewed in [5]) likely due to priming of STAT levels [89], TLRs and RLRs [5], and the inflammasome [90, 91]. Indeed, we and others have demonstrated that type I IFNs prime the basal levels of hundreds of IRGs, many of which play central roles in signaling by other systems [42, 92]. Important among these are negative regulators such as SOCS1, which are not only rapidly and strongly IFN inducible but play important roles in dampening responses to type I and type II IFNs, other cytokines and TLRs [93–95]. Indeed neonatal mice deficient in SOCS 1 die from multi-organ inflammation in the absence of SOCS1 suppression of type I [94] and type II IFN signaling [93].

Special Case Study of IFNβ

As discussed above, IFNβ is different from other type I IFNs in being the only one induced by LPS, thus playing a central role in response to bacteria [92, 96]. In addition, the promoter of IFNβ is unusual in having AP1 sites that can be activated by the fos/jun and MAP kinase pathway. This pathway is activated during macrophage development in response to M-CSF and in osteoclast development in response to RANK Ligand [11]. The inhibitory effect of on IFNβ on the proliferation of these myeloid cells may be important in the regulation of pathogen responses. In addition to selective production of IFNβ relative to other type I IFNs, it has a higher binding affinity to receptors and is more potent than the members of the IFNα family in antiproliferative assays on certain cell types. It is a singularly effective therapeutically in multiple sclerosis [97]. However, until recently, there has been no mechanistic explanation for differential activities of IFNb relative to other type I IFNs..

De Weerd et al. [97] demonstrated that IFNβ but not IFNα formed a complex with IFNAR1 in the absence of IFNAR2. Crystallization of the IFNβ:IFNAR1 complex showed extensive contacts of this IFN with the receptor over a much larger surface area in that crystal structure than any potential IFNα:IFNAR1 [98]. Further studies of *Ifnar2* null cells showed that while the binding of IFNβ to IFNAR1 did not induce canonical STAT signaling as expected, there were signals transduced. Approximately 230 genes were induced by this novel IFNβ:IFNAR1 signaling axis by an uncharacterized pathway. Induced genes included several such as TREM1, TREML4, TGM2, and CCL2, which had known roles in the response to sepsis. Using an in vivo murine model of LPS-induced septic shock, it was demonstrated that this unique IFNβ:IFNAR1 signaling axis was important in mediating the previously described IFN-mediated toxicity.

Specifically, this study shows molecular mechanisms whereby IFNβ can transduce specific signals with pathophysiological importance. In general terms it opens the door for discovering previously elusive selective actions of other type I IFNs by differential interaction with IFNAR1 and IFNAR2. Similarly, cells may regulate the response to type I IFNs by differential regulation of the cell surface expression of IFNAR1 and IFNAR2.

Special Case Study of IFNε

Recently, the function of a specialized type I IFN was reported. IFNε was characterized as a type I IFN based on sequence homology, the location of the gene in the type I IFN gene locus on human chromosome 9p (and syntenic murine chromosome 16) and its signaling through IFNAR1 and IFNAR2 [42, 99]. Recombinant IFNe protein induced “classical” IRGs like other type I IFNs and this signaling was abrogated in cells from *Ifnar1* or *Ifnar2* deficient mice [42]. However, the expression patterns and regulation of this gene showed unique features. Unlike other type I IFNs, it was not pathogen inducible and was constitutively expressed. This constitutive expression was most notable in the female reproductive tract (FRT). Also unlike other type I IFNs, its expression was regulated by hormones: stimulated by estrogen and repressed by progesterone. Accordingly, its expression fluctuated during the female cycle, was dramatically reduced at the time of embryo implantation in the mouse, and was reduced to virtually undetectable in post-menopausal women, when estrogen levels decline. The in vivo functional importance of IFNε was determined in IFNε-deficient mice. These mice were more susceptible to viral infection with HSV and bacterial infection with *C. muridarum*. The constitutive production of IFNε in the epithelial cells of the endometrium maintained the basal expression of many IRGs including those involved in pathogen defense (Mx, ISG 15, IRGM1) and PRR sensing and primary signaling (IRF7). This priming of the innate immune response by constitutive IFNε ensures protection of the FRT mucosa from early stages of viral and bacterial infection. Furthermore, the absence of IFNε also restricted bacterial clearance, consistent with the continued production of this protective cytokine before and throughout the course of infection since this pathogen did not modulate IFNε expression in vivo. The levels of NK cells, which have been shown to aid clearance of pathogen, correlated with the levels of IFNε: administration of recombinant mu IFNε to IFNε-null mice restored the depleted levels of NK cells and decreased the number of bacteria recovered 3 days post infection. Interestingly, IFNε is the only type I IFN that protects the FRT from *Chlamydia* infection. *Ifnar1* deficient mice show less severe disease; indicating the exacerbation of disease by production of (presumably conventional, α/β) type I IFNs; shown by adoptive transfer experiments to be acting on CD8 T cells driving disease pathogenesis [75]. This is similar to infections with other intracellular bacteria such as *L. monocytogenes*, *F. tularensis*, *M. tuberculosis*, in which disease pathology is exacerbated by type I IFNs (refer above).

Thus, the actions of IFNε in protecting the FRT highlight several general principles that might be applicable to IFN anti-pathogen strategies in general: (1) it is a direct example of how regulating expression in a particular way can achieve specific and functional protection; (2) it shows a specific adaptation of the innate immune response to suit organ-specific requirements of host defense; and (3) it shows how compartmentalization of an IFN response can achieve opposite outcomes—epithelial production of IFNε is protective, whereas mucosal production of conventional IFNs exacerbates disease through their action on immune cells.

Concluding Remarks

The type I IFNs have pleiotropic effects on host defense due to their ability to regulate the parenchymal cells under attack by infectious agents or the innate and adaptive immune cells that traffic to and from the site of infection. While we have made considerable advances in understanding mechanisms of signal transduction via the IFNARs, JAK/STAT and other signal transduction pathways, we are only just beginning to understand the cell context and temporal specificities of type I IFN signaling and responses. This is manifest in the different transcription profiles identified in different cell types responding to IFNs, which represents only a part of the available repertoire of IRGs that encode the effector molecules. Understanding and harnessing the specificity of the response will make inroads into understanding and dealing with the current and emerging threats posed by bacteria and other pathogens.

References

1. Pestka S, Krause CD, Walter MR (2004) Interferons, interferon-like cytokines, and their receptors. Immunol Rev 202:8–32
2. de Weerd NA, Samarajiwa SA, Hertzog PJ (2007) Type I interferon receptors: biochemistry and biological functions. J Biol Chem 282:20053–20057
3. Schroder K, Hertzog PJ, Ravasi T, Hume DA (2004) Interferon-gamma: an overview of signals, mechanisms and functions. J Leukoc Biol 75:163–189
4. Kotenko SV (2011) IFN-lambdas. Curr Opin Immunol 23:583–590
5. Hertzog PJ, Williams BR (2013) Fine tuning type I interferon responses. Cytokine Growth Factor Rev 24:217–225
6. Nagano Y, Kojima Y (1954) Pouvoir immunisant du virus vaccinal inactivé par des rayons ultraviolets. Comptes rendus des séances de la Société de biologie et de ses filiales 148: 1700–1702
7. Isaacs A, Lindenmann J (1957) Virus interference. I. The interferon. Proc R Soc Lond B Biol Sci 147:258–267
8. Levy DE, Marie IJ, Durbin JE (2011) Induction and function of type I and III interferon in response to viral infection. Curr Opin Virol 1:476–486
9. Monroe KM, McWhirter SM, Vance RE (2010) Induction of type I interferons by bacteria. Cell Microbiol 12:881–890
10. Hamilton JA, Whitty GA, Kola I, Hertzog PJ (1996) Endogenous IFN-alpha beta suppresses colony-stimulating factor (CSF)-1-stimulated macrophage DNA synthesis and mediates inhibitory effects of lipopolysaccharide and TNF-alpha. J Immunol 156:2553–2557
11. Takayanagi H, Kim S, Taniguchi T (2002) Signaling crosstalk between RANKL and interferons in osteoclast differentiation. Arthritis Res 4(Suppl 3):S227–S232
12. Swann JB, Hayakawa Y, Zerafa N, Sheehan KC, Scott B, Schreiber RD, Hertzog P, Smyth MJ (2007) Type I IFN contributes to NK cell homeostasis, activation, and antitumor function. J Immunol 178:7540–7549
13. Bidwell BN, Slaney CY, Withana NP, Forster S, Cao Y, Loi S, Andrews D, Mikeska T, Mangan NE, Samarajiwa SA, de Weerd NA, Gould J, Argani P, Moller A, Smyth MJ, Anderson RL, Hertzog PJ, Parker BS (2012) Silencing of Irf7 pathways in breast cancer cells promotes bone metastasis through immune escape. Nat Med 18:1224–1231
14. Hertzog PJ, O'Neill LA, Hamilton JA (2003) The interferon in TLR signaling: more than just antiviral. Trends Immunol 24:534–539
15. Noppert SJ, Fitzgerald KA, Hertzog PJ (2007) The role of type I interferons in TLR responses. Immunol Cell Biol 85:446–457
16. Kawai T, Akira S (2010) The role of pattern-recognition receptors in innate immunity: update on Toll-like receptors. Nat Immunol 11:373–384
17. Tamura T, Yanai H, Savitsky D, Taniguchi T (2008) The IRF family transcription factors in immunity and oncogenesis. Annu Rev Immunol 26:535–584

18. Barbalat R, Lau L, Locksley RM, Barton GM (2009) Toll-like receptor 2 on inflammatory monocytes induces type I interferon in response to viral but not bacterial ligands. Nat Immunol 10:1200–1207
19. Weiss G, Maaetoft-Udsen K, Stifter SA, Hertzog P, Goriely S, Thomsen AR, Paludan SR, Frokiaer H (2012) MyD88 drives the IFN-beta response to *Lactobacillus acidophilus* in dendritic cells through a mechanism involving IRF1, IRF3, and IRF7. J Immunol 189: 2860–2868
20. Parker D, Prince A (2012) *Staphylococcus aureus* induces type I IFN signaling in dendritic cells via TLR9. J Immunol 189:4040–4046
21. Mancuso G, Gambuzza M, Midiri A, Biondo C, Papasergi S, Akira S, Teti G, Beninati C (2009) Bacterial recognition by TLR7 in the lysosomes of conventional dendritic cells. Nat Immunol 10:587–594
22. Kato H, Sato S, Yoneyama M, Yamamoto M, Uematsu S, Matsui K, Tsujimura T, Takeda K, Fujita T, Takeuchi O, Akira S (2005) Cell type-specific involvement of RIG-I in antiviral response. Immunity 23:19–28
23. Kawai T, Takahashi K, Sato S, Coban C, Kumar H, Kato H, Ishii KJ, Takeuchi O, Akira S (2005) IPS-1, an adaptor triggering RIG-I- and Mda5-mediated type I interferon induction. Nat Immunol 6:981–988
24. Seth RB, Sun L, Ea CK, Chen ZJ (2005) Identification and characterization of MAVS, a mitochondrial antiviral signaling protein that activates NF-kappaB and IRF 3. Cell 122:669–682
25. Ishikawa H, Ma Z, Barber GN (2009) STING regulates intracellular DNA-mediated, type I interferon-dependent innate immunity. Nature 461:788–792
26. Barber GN (2014) STING-dependent cytosolic DNA sensing pathways. Trends Immunol 35:88–93
27. Woodward JJ, Iavarone AT, Portnoy DA (2010) c-di-AMP secreted by intracellular *Listeria monocytogenes* activates a host type I interferon response. Science 328:1703–1705
28. Bowie AG (2012) Innate sensing of bacterial cyclic dinucleotides: more than just STING. Nat Immunol 13:1137–1139
29. Parvatiyar K, Zhang Z, Teles RM, Ouyang S, Jiang Y, Iyer SS, Zaver SA, Schenk M, Zeng S, Zhong W, Liu ZJ, Modlin RL, Liu YJ, Cheng G (2012) The helicase DDX41 recognizes the bacterial secondary messengers cyclic di-GMP and cyclic di-AMP to activate a type I interferon immune response. Nat Immunol 13:1155–1161
30. Prantner D, Darville T, Nagarajan UM (2010) Stimulator of IFN gene is critical for induction of IFN-beta during *Chlamydia muridarum* infection. J Immunol 184:2551–2560
31. Manzanillo PS, Shiloh MU, Portnoy DA, Cox JS (2012) *Mycobacterium tuberculosis* activates the DNA-dependent cytosolic surveillance pathway within macrophages. Cell Host Microbe 11:469–480
32. Lippmann J, Muller HC, Naujoks J, Tabeling C, Shin S, Witzenrath M, Hellwig K, Kirschning CJ, Taylor GA, Barchet W, Bauer S, Suttorp N, Roy CR, Opitz B (2011) Dissection of a type I interferon pathway in controlling bacterial intracellular infection in mice. Cell Microbiol 13:1668–1682
33. Takaoka A, Wang Z, Choi MK, Yanai H, Negishi H, Ban T, Lu Y, Miyagishi M, Kodama T, Honda K, Ohba Y, Taniguchi T (2007) DAI (DLM-1/ZBP1) is a cytosolic DNA sensor and an activator of innate immune response. Nature 448:501–505
34. Parker D, Martin FJ, Soong G, Harfenist BS, Aguilar JL, Ratner AJ, Fitzgerald KA, Schindler C, Prince A (2011) *Streptococcus pneumoniae* DNA initiates type I interferon signaling in the respiratory tract. MBio 2:e00016-00011
35. Chiu YH, Macmillan JB, Chen ZJ (2009) RNA polymerase III detects cytosolic DNA and induces type I interferons through the RIG-I pathway. Cell 138:576–591
36. Henry T, Brotcke A, Weiss DS, Thompson LJ, Monack DM (2007) Type I interferon signaling is required for activation of the inflammasome during *Francisella* infection. J Exp Med 204:987–994
37. Henry T, Monack DM (2007) Activation of the inflammasome upon *Francisella tularensis* infection: interplay of innate immune pathways and virulence factors. Cell Microbiol 9: 2543–2551

38. Kim S, Bauernfeind F, Ablasser A, Hartmann G, Fitzgerald KA, Latz E, Hornung V (2010) *Listeria monocytogenes* is sensed by the NLRP3 and AIM2 inflammasome. Eur J Immunol 40:1545–1551
39. Pandey AK, Yang Y, Jiang Z, Fortune SM, Coulombe F, Behr MA, Fitzgerald KA, Sassetti CM, Kelliher MA (2009) NOD2, RIP2 and IRF5 play a critical role in the type I interferon response to *Mycobacterium tuberculosis*. PLoS Pathog 5:e1000500
40. Park JH, Kim YG, McDonald C, Kanneganti TD, Hasegawa M, Body-Malapel M, Inohara N, Nunez G (2007) RICK/RIP2 mediates innate immune responses induced through Nod1 and Nod2 but not TLRs. J Immunol 178:2380–2386
41. Parker D, Planet PJ, Soong G, Narechania A, Prince A (2014) Induction of type I interferon signaling determines the relative pathogenicity of *Staphylococcus aureus* strains. PLoS Pathog 10:e1003951
42. Fung KY, Mangan NE, Cumming H, Horvat JC, Mayall JR, Stifter SA, De Weerd N, Roisman LC, Rossjohn J, Robertson SA, Schjenken JE, Parker B, Gargett CE, Nguyen HP, Carr DJ, Hansbro PM, Hertzog PJ (2013) Interferon-epsilon protects the female reproductive tract from viral and bacterial infection. Science 339:1088–1092
43. Parker D, Prince A (2011) Type I interferon response to extracellular bacteria in the airway epithelium. Trends Immunol 32:582–588
44. Hwang SY, Hertzog PJ, Holland KA, Sumarsono SH, Tymms MJ, Hamilton JA, Whitty G, Bertoncello I, Kola I (1995) A null mutation in the gene encoding a type I interferon receptor component eliminates antiproliferative and antiviral responses to interferons alpha and beta and alters macrophage responses. Proc Natl Acad Sci U S A 92:11284–11288
45. Muller U, Steinhoff U, Reis LF, Hemmi S, Pavlovic J, Zinkernagel RM, Aguet M (1994) Functional role of type I and type II interferons in antiviral defense. Science 264:1918–1921
46. Novick D, Cohen B, Rubinstein M (1994) The human interferon alpha/beta receptor: characterization and molecular cloning. Cell 77:391–400
47. Lutfalla G, Holland SJ, Cinato E, Monneron D, Reboul J, Rogers NC, Smith JM, Stark GR, Gardiner K, Mogensen KE et al (1995) Mutant U5A cells are complemented by an interferon-alpha beta receptor subunit generated by alternative processing of a new member of a cytokine receptor gene cluster. EMBO J 14:5100–5108
48. Owczarek CM, Hwang SY, Holland KA, Gulluyan LM, Tavaria M, Weaver B, Reich NC, Kola I, Hertzog PJ (1997) Cloning and characterization of soluble and transmembrane isoforms of a novel component of the murine type I interferon receptor, IFNAR 2. J Biol Chem 272: 23865–23870
49. Hardy MP, Owczarek CM, Trajanovska S, Liu X, Kola I, Hertzog PJ (2001) The soluble murine type I interferon receptor Ifnar-2 is present in serum, is independently regulated, and has both agonistic and antagonistic properties. Blood 97:473–482
50. Ruwanpura SM, McLeod L, Brooks GD, Bozinovski S, Vlahos R, Longano A, Bardin PG, Anderson GP, Jenkins BJ (2014) IL-6/Stat3-driven pulmonary inflammation, but not emphysema, is dependent on interleukin-17A in mice. Respirology 19:419–427
51. Samarajiwa SA, Mangan NE, Hardy MP, Najdovska M, Dubach D, Braniff SJ, Owczarek CM, Hertzog PJ (2014) Soluble IFN receptor potentiates in vivo type I IFN signaling and exacerbates TLR4-mediated septic shock. J Immunol 192:4425–4435
52. Domanski P, Fish E, Nadeau OW, Witte M, Platanias LC, Yan H, Krolewski J, Pitha P, Colamonici OR (1997) A region of the beta subunit of the interferon alpha receptor different from box 1 interacts with Jak1 and is sufficient to activate the Jak-Stat pathway and induce an antiviral state. J Biol Chem 272:26388–26393
53. Yan H, Krishnan K, Lim JT, Contillo LG, Krolewski JJ (1996) Molecular characterization of an alpha interferon receptor 1 subunit (IFNaR1) domain required for TYK2 binding and signal transduction. Mol Cell Biol 16:2074–2082
54. Stark GR, Darnell JE Jr (2012) The JAK-STAT pathway at twenty. Immunity 36:503–514
55. Piganis RA, De Weerd NA, Gould JA, Schindler CW, Mansell A, Nicholson SE, Hertzog PJ (2011) Suppressor of cytokine signaling (SOCS) 1 inhibits type I interferon (IFN) signaling

via the interferon alpha receptor (IFNAR1)-associated tyrosine kinase Tyk2. J Biol Chem 286:33811–33818
56. Zhao W, Lee C, Piganis R, Plumlee C, de Weerd N, Hertzog PJ, Schindler C (2008) A conserved IFN-alpha receptor tyrosine motif directs the biological response to type I IFNs. J Immunol 180:5483–5489
57. Platanias LC (2005) Mechanisms of type-I- and type-II-interferon-mediated signalling. Nat Rev Immunol 5:375–386
58. van Boxel-Dezaire AH, Rani MR, Stark GR (2006) Complex modulation of cell type-specific signaling in response to type I interferons. Immunity 25:361–372
59. O'Shea JJ, Plenge R (2012) JAK and STAT signaling molecules in immunoregulation and immune-mediated disease. Immunity 36:542–550
60. Ragimbeau J, Dondi E, Alcover A, Eid P, Uze G, Pellegrini S (2003) The tyrosine kinase Tyk2 controls IFNAR1 cell surface expression. EMBO J 22:537–547
61. Ramana CV, Gil MP, Han Y, Ransohoff RM, Schreiber RD, Stark GR (2001) Stat1-independent regulation of gene expression in response to IFN-gamma. Proc Natl Acad Sci U S A 98:6674–6679
62. Yang CH, Murti A, Pfeffer SR, Kim JG, Donner DB, Pfeffer LM (2001) Interferon alpha/beta promotes cell survival by activating nuclear factor kappa B through phosphatidylinositol 3-kinase and Akt. J Biol Chem 276:13756–13761
63. Joshi S, Kaur S, Redig AJ, Goldsborough K, David K, Ueda T, Watanabe-Fukunaga R, Baker DP, Fish EN, Fukunaga R, Platanias LC (2009) Type I interferon (IFN)-dependent activation of Mnk1 and its role in the generation of growth inhibitory responses. Proc Natl Acad Sci U S A 106:12097–12102
64. Kaur S, Sassano A, Dolniak B, Joshi S, Majchrzak-Kita B, Baker DP, Hay N, Fish EN, Platanias LC (2008) Role of the Akt pathway in mRNA translation of interferon-stimulated genes. Proc Natl Acad Sci U S A 105:4808–4813
65. Kaur S, Sassano A, Majchrzak-Kita B, Baker DP, Su B, Fish EN, Platanias LC (2012) Regulatory effects of mTORC2 complexes in type I IFN signaling and in the generation of IFN responses. Proc Natl Acad Sci U S A 109:7723–7728
66. Petricoin EF 3rd, Ito S, Williams BL, Audet S, Stancato LF, Gamero A, Clouse K, Grimley P, Weiss A, Beeler J, Finbloom DS, Shores EW, Abraham R, Larner AC (1997) Antiproliferative action of interferon-alpha requires components of T-cell-receptor signalling. Nature 390:629–632
67. Samarajiwa SA, Forster S, Auchettl K, Hertzog PJ (2009) INTERFEROME: the database of interferon regulated genes. Nucleic Acids Res 37:D852–D857
68. Rusinova I, Forster S, Yu S, Kannan A, Masse M, Cumming H, Chapman R, Hertzog PJ (2013) Interferome v2.0: an updated database of annotated interferon-regulated genes. Nucleic Acids Res 41:D1040–D1046
69. Hertzog P, Forster S, Samarajiwa S (2011) Systems biology of interferon responses. J Interferon Cytokine Res 31:5–11
70. Harman AN, Lai J, Turville S, Samarajiwa S, Gray L, Marsden V, Mercier SK, Jones K, Nasr N, Rustagi A, Cumming H, Donaghy H, Mak J, Gale M Jr, Churchill M, Hertzog P, Cunningham AL (2011) HIV infection of dendritic cells subverts the IFN induction pathway via IRF-1 and inhibits type 1 IFN production. Blood 118:298–308
71. Berry MP, Blankley S, Graham CM, Bloom CI, O'Garra A (2013) Systems approaches to studying the immune response in tuberculosis. Curr Opin Immunol 25:579–587
72. Schoggins JW (2014) Interferon-stimulated genes: roles in viral pathogenesis. Curr Opin Virol 6C:40–46
73. Decker T, Muller M, Stockinger S (2005) The yin and yang of type I interferon activity in bacterial infection. Nat Rev Immunol 5:675–687
74. Gonzalez-Navajas JM, Lee J, David M, Raz E (2012) Immunomodulatory functions of type I interferons. Nat Rev Immunol 12:125–135

75. Nagarajan UM, Prantner D, Sikes JD, Andrews CW Jr, Goodwin AM, Nagarajan S, Darville T (2008) Type I interferon signaling exacerbates *Chlamydia muridarum* genital infection in a murine model. Infect Immun 76:4642–4648
76. de Almeida LA, Carvalho NB, Oliveira FS, Lacerda TL, Vasconcelos AC, Nogueira L, Bafica A, Silva AM, Oliveira SC (2011) MyD88 and STING signaling pathways are required for IRF3-mediated IFN-beta induction in response to *Brucella abortus* infection. PLoS One 6:e23135
77. Lizak M, Yarovinsky TO (2012) Phospholipid scramblase 1 mediates type I interferon-induced protection against staphylococcal alpha-toxin. Cell Host Microbe 11:70–80
78. Taylor GA, Feng CG, Sher A (2004) p47 GTPases: regulators of immunity to intracellular pathogens. Nat Rev Immunol 4:100–109
79. Hervas-Stubbs S, Perez-Gracia JL, Rouzaut A, Sanmamed MF, Le Bon A, Melero I (2011) Direct effects of type I interferons on cells of the immune system. Clin Cancer Res 17: 2619–2627
80. Carrigan SO, Junkins R, Yang YJ, Macneil A, Richardson C, Johnston B, Lin TJ (2010) IFN regulatory factor 3 contributes to the host response during *Pseudomonas aeruginosa* lung infection in mice. J Immunol 185:3602–3609
81. Power MR, Li B, Yamamoto M, Akira S, Lin TJ (2007) A role of Toll-IL-1 receptor domain-containing adaptor-inducing IFN-beta in the host response to *Pseudomonas aeruginosa* lung infection in mice. J Immunol 178:3170–3176
82. Chang EY, Guo B, Doyle SE, Cheng G (2007) Cutting edge: involvement of the type I IFN production and signaling pathway in lipopolysaccharide-induced IL-10 production. J Immunol 178:6705–6709
83. Guo B, Chang EY, Cheng G (2008) The type I IFN induction pathway constrains Th17-mediated autoimmune inflammation in mice. J Clin Invest 118:1680–1690
84. Carrero JA, Unanue ER (2006) Lymphocyte apoptosis as an immune subversion strategy of microbial pathogens. Trends Immunol 27:497–503
85. Stephanou A, Brar BK, Knight RA, Latchman DS (2000) Opposing actions of STAT-1 and STAT-3 on the Bcl-2 and Bcl-x promoters. Cell Death Differ 7:329–330
86. Broz P, Monack DM (2013) Noncanonical inflammasomes: caspase-11 activation and effector mechanisms. PLoS Pathog 9:e1003144
87. Robinson N, McComb S, Mulligan R, Dudani R, Krishnan L, Sad S (2012) Type I interferon induces necroptosis in macrophages during infection with *Salmonella enterica* serovar *Typhimurium*. Nat Immunol 13:954–962
88. Ivashkiv LB, Donlin LT (2014) Regulation of type I interferon responses. Nat Rev Immunol 14:36–49
89. Gough DJ, Messina NL, Hii L, Gould JA, Sabapathy K, Robertson AP, Trapani JA, Levy DE, Hertzog PJ, Clarke CJ, Johnstone RW (2010) Functional crosstalk between type I and II interferon through the regulated expression of STAT1. PLoS Biol 8:e1000361
90. Guarda G, Braun M, Staehli F, Tardivel A, Mattmann C, Forster I, Farlik M, Decker T, Du Pasquier RA, Romero P, Tschopp J (2011) Type I interferon inhibits interleukin-1 production and inflammasome activation. Immunity 34:213–223
91. Malireddi RK, Kanneganti TD (2013) Role of type I interferons in inflammasome activation, cell death, and disease during microbial infection. Front Cell Infect Microbiol 3:77
92. Thomas KE, Galligan CL, Newman RD, Fish EN, Vogel SN (2006) Contribution of interferon-beta to the murine macrophage response to the toll-like receptor 4 agonist, lipopolysaccharide. J Biol Chem 281:31119–31130
93. Alexander WS, Starr R, Fenner JE, Scott CL, Handman E, Sprigg NS, Corbin JE, Cornish AL, Darwiche R, Owczarek CM, Kay TW, Nicola NA, Hertzog PJ, Metcalf D, Hilton DJ (1999) SOCS1 is a critical inhibitor of interferon gamma signaling and prevents the potentially fatal neonatal actions of this cytokine. Cell 98:597–608
94. Fenner JE, Starr R, Cornish AL, Zhang JG, Metcalf D, Schreiber RD, Sheehan K, Hilton DJ, Alexander WS, Hertzog PJ (2006) Suppressor of cytokine signaling 1 regulates the immune

response to infection by a unique inhibition of type I interferon activity. Nat Immunol 7: 33–39
95. Mansell A, Smith R, Doyle SL, Gray P, Fenner JE, Crack PJ, Nicholson SE, Hilton DJ, O'Neill LA, Hertzog PJ (2006) Suppressor of cytokine signaling 1 negatively regulates Toll-like receptor signaling by mediating Mal degradation. Nat Immunol 7:148–155
96. Karaghiosoff M, Steinborn R, Kovarik P, Kriegshauser G, Baccarini M, Donabauer B, Reichart U, Kolbe T, Bogdan C, Leanderson T, Levy D, Decker T, Muller M (2003) Central role for type I interferons and Tyk2 in lipopolysaccharide-induced endotoxin shock. Nat Immunol 4: 471–477
97. de Weerd NA, Vivian JP, Nguyen TK, Mangan NE, Gould JA, Braniff SJ, Zaker-Tabrizi L, Fung KY, Forster SC, Beddoe T, Reid HH, Rossjohn J, Hertzog PJ (2013) Structural basis of a unique interferon-beta signaling axis mediated via the receptor IFNAR1. Nat Immunol 14: 901–907
98. Thomas C, Moraga I, Levin D, Krutzik PO, Podoplelova Y, Trejo A, Lee C, Yarden G, Vleck SE, Glenn JS, Nolan GP, Piehler J, Schreiber G, Garcia KC (2011) Structural linkage between ligand discrimination and receptor activation by type I interferons. Cell 146:621–632
99. Hardy MP, Owczarek CM, Jermiin LS, Ejdeback M, Hertzog PJ (2004) Characterization of the type I interferon locus and identification of novel genes. Genomics 84:331–345

Induction and Consequences of the Type I IFN Response to *Listeria monocytogenes*

Emily M. Eshleman and Laurel L. Lenz

Introduction

Listeria monocytogenes is a Gram-positive, facultative intracellular bacterium that causes foodborne illnesses in animals and humans. *L. monocytogenes* is the causative agent of listeriosis, a life-threatening systemic infection that primarily affects aged or immune compromised individuals and pregnant women. Clinical features of *L. monocytogenes* infection range in severity from gastroenteritis to septicemia and meningitis. When infecting pregnant individuals, *L. monocytogenes* also causes abortions, still births, and neonatal meningitis. The incidence of listeriosis is low, but the mortality rate is high. Hence, *L. monocytogenes* remains a leading cause of death from foodborne illness within the USA. For example, in 2011 a *L. monocytogenes* outbreak associated with cantaloupes infected 147 individuals with 33 deaths, for a mortality rate of 22 % [1].

L. monocytogenes gains entry into a wide variety of mammalian cells, both hematopoietic and non-hematopoietic, by phagocytosis or clathrin mediated uptake [2–4]. The bacterium usually does not replicate within phagosomes or vacuolar compartments but instead escapes these compartments to grow in the cell cytosol. A major bacterial virulence factor required for phagosomal escape is the pore-forming toxin listeriolysin O (LLO), encoded by the *hly* gene. LLO is secreted and active preferentially under acidic conditions found in maturing phagosomes, where

E.M. Eshleman
Integrated Department of Immunology, University of Colorado School of Medicine, Aurora, CO 80045, USA

L.L. Lenz (✉)
Integrated Department of Immunology, University of Colorado School of Medicine, Aurora, CO 80045, USA

Integrated Department of Immunology, National Jewish Health, Denver, CO 80206, USA
e-mail: LenzL@NJHealth.org

© Springer International Publishing Switzerland 2014

D. Parker (ed.), *Bacterial Activation of Type I Interferons*,
DOI 10.1007/978-3-319-09498-4_2

it destroys the phagosomal membrane with additional contributions by two bacterial phospholipases [5]. The exact mechanism of phagosomal escape is still under debate. However, *L. monocytogenes* strains with mutation of *hly* or otherwise deficient in LLO are attenuated and fail to escape acidified phagosomes [6]. *L. monocytogenes* strains that invade into the cytosol trigger CD8+ T cell responses and long-lasting protective immunity, while LLO-deficient strains are poor at eliciting CD8+ T cell responses and protective immunity [7].

Following systemic infection of mice, *L. monocytogenes* primarily localizes to the liver and spleen. The bacteria are rapidly phagocytosed by resident macrophages and dendritic cells (DC) within these tissues. Some of the phagocytosed bacteria escape phagosomes and replicate within these cells. In response to *L. monocytogenes*, phagocytes produce pro-inflammatory cytokines such as TNFα and type I interferons (IFN). Type I IFNs have long been associated with effective anti-viral immunity, but their role during bacterial infections is less clear. During infections by *L. monocytogenes*, *Mycobacterium tuberculosis*, and several other bacteria type I IFN are detrimental to the host. A better understanding of how type I IFN responses are regulated during *L. monocytogenes* infection thus has potential impact for treatment of bacterial infections. Though much has been learned in this regard, the detailed mechanisms for induction of these cytokines (abbreviated IFN-α/β) are still being unraveled. The goal of this chapter is to summarize the current state of research in this area. We outline the pattern recognition receptors (PRRs) and signaling pathways involved in the production of type I IFNs during a *L. monocytogenes* infection and the biological effects their production has on the host. Pathways known to be important for induction of type I IFN within *L. monocytogenes*-infected phagocytes are diagrammed in Fig. 1.

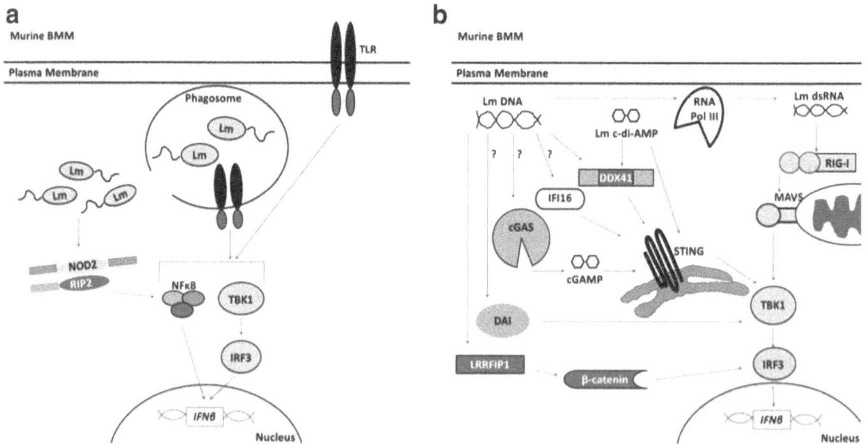

Fig. 1 Mammalian sensing of *L. monocytogenes* microbial components leading to induction of type I IFNs. (**a**) TLR and NOD pathways commonly recognize *L. monocytogenes* cell wall and envelope moieties. These pathways have not been shown to be required, but may augment IFN-β production. (**b**) Nucleic acid sensing pathways are known to induce type I IFNs by *L. monocytogenes* secretion of DNA, RNA, and cyclic di-nucleotides. While many of these pathways have been verified by direct recognition of *L. monocytogenes* nucleic acids, question marks (?) indicate potential but unconfirmed *L. monocytogenes* DNA sensors

IFN Regulatory Factor 3 Is Crucial for Type I IFN Responses During *L. monocytogenes* Infection

Members of the IFN regulatory factor (IRF) family of transcription factors regulate type I IFN production during viral infections and in response to other inflammatory stimuli. IRF3 in particular acts as an early factor regulating the type I IFN response. In resting cells, IRF3 is found in an inactivated state within the cytoplasm [8]. Phosphorylation on serine residues near the C-terminus of IRF3 enables it to dimerize and form complexes with CBP/p300, and to translocate to the nucleus where it can bind promoter regions of *Ifnb* and other genes. IRF3 thus helps initiate *Ifnb* transcription and subsequent secretion of IFN-β [8]. Once produced, IFN-β mediates autocrine and paracrine signaling through the IFN-α/β receptor (IFNAR). Such signaling activates transcriptional complexes involving STAT1, STAT2, IRF7, and IRF9. These complexes bind promoters to regulate expression of diverse interferon regulated genes (IRGs), including those encoding other type I IFNs (e.g. IFNα proteins). Thus, IRF3 activation directly or indirectly triggers production of multiple type I IFN proteins.

IRF3 is involved in IFN-β production during *L. monocytogenes* infection of macrophages. Specifically, infected murine bone marrow derived macrophages (BMDMs) showed significant nuclear localization of IRF3 at 4 h after infection [9]. Unlike wild-type BMMs, BMMs derived from IRF3-deficient mice also failed to induce expression of IFN-β upon infection by *L. monocytogenes* [9]. Studies with C57Bl/6ByJ mice also indicated an important role for IRF3 in the response to *L. monocytogenes*. BMDMs from this inbred sub strain of C57Bl/6 mice transcribed ~100-fold lower *Ifnb* mRNA upon *L. monocytogenes* infection [10]. Consistent with the reduced type I IFN response, these mice also showed significantly increased resistance to challenge with a lethal infection dose. The defect in type I IFN production mapped to a single A-T mutation found to be important for efficient splicing of *Irf3*. This mutation resulted in reduced IRF3 protein levels that correlated with the reduced type I IFN synthesis [10]. Subsequent studies from several other groups have independently confirmed the importance of IRF3 in the induction of type I IFNs by *L. monocytogenes* [9, 11–14].

TNFR-Associated NF-κB Kinase- Binding Kinase 1 (TBK1) is Crucial for Type I IFN Responses During *L. monocytogenes* Infection

The phosphorylation of IRF3 and stimulation of IFN-β production during viral infections or stimulation of cells with dsRNA requires two serine kinases, TNFR-associated NF-κB kinase (TANK)-binding kinase 1 (TBK1) and I-κB kinase ε (IKKε) [15–18]. TBK1 is an ubiquitously expressed member of the IKK protein kinase family that can associate with IKKε and TANK to regulate NF-κB activation and expression of several proinflammatory cytokines [19]. Knockdown of either IKKε or TBK1 abolishes the production of IFN-β in response to dsRNA stimulation, suggesting a non-redundant role for these two kinases [16]. Evidence that TBK1 plays a role in IRF3 activation during a *L. monocytogenes* infection comes from experiments with infected murine embryonic fibroblasts (MEF) from TBK1 sufficient and deficient littermates. Unlike control MEFs, *Tbk1*$^{-/-}$ MEF showed no nuclear translocation of IRF3 and no production of IFN-β [11]. In contrast, infection-induced nuclear localization of the p65 NFκB subunit was not affected by TBK1 deficiency, suggesting a specific requirement for TBK1 in IRF3 activation [11]. Additional evidence that TBK1 promotes IRF3 nuclear translocation and type I IFN synthesis during an infection with *L. monocytogenes* comes from studies with BMDMs lacking both TBK1 and TNFR1. The double knockout cells were used as TBK1 deletion causes embryonic lethality in TNF-responsive mice. IFN-β production by the *Tbk1*$^{-/-}$*Tnfr1*$^{-/-}$ BMDMs was drastically, but not completely, reduced [11]. These results demonstrate that TBK1 is important but also argue there may be some functional overlap between TBK1 and IKKε in IRF3 activation during *L. monocytogenes* challenge [11, 16].

Toll-Like Receptors Recognize *L. monocytogenes* and in Some Situations May Contribute to Type I IFN Production

The Toll-like receptor (TLR) family of transmembrane receptors recognize molecular patterns associated with bacteria and viruses (PAMPs). Ligation of various TLRs by microbial products initiates signaling pathways involving NFκB, MAPK, and in several cases IRFs [20]. Thus, stimulation of TLRs can result in the production of proinflammatory cytokines and in some cases type I IFNs. The extracellular regions of TLRs contain leucine rich repeats (LRR) that mediate ligand binding, while their cytosolic regions contain Toll/IL-1 receptor (TIR) domains that interacts with other TIR containing adaptor proteins. Notably, TIR domains in TLRs recruit signaling adapters myeloid differentiation primary response gene 88 (MyD88) and/or TIR domain containing adapter inducing IFN-β (TRIF) [20]. This latter factor associates with TBK1 to ultimately stimulate IRF3 activation and IFN-β production.

Work with mouse cells has shown that several TLRs are capable of detecting *L. monocytogenes* products. In some cases, such recognition might conceivably contribute to the induction of type I IFNs. For example, TLRs 2, 3, and 4 have been shown to recruit TRIF to activate TBK1, IRF3, and production of IFN-α/β [18, 21]. TLR4 is best known as the receptor for lipopolysaccharide (LPS), which is produced exclusively by Gram-negative bacteria. However, TLR4 can also reportedly recognize lipoteichoic acids present in the cell envelope of *L. monocytogenes* and other Gram-positive bacteria [22]. Nevertheless, TLR4 expression was not required for nuclear translocation of IRF3 or type I IFN production by *L. monocytogenes-infected* BMMs [11]. TLR4 deficiency also failed to reduce IFN-β production by

L. monocytogenes-infected peritoneal macrophages [13]. TLR3 stimuli are well known to elicit type I IFN production. However, this TLR recognizes double-stranded RNA present in certain viral particles or produced during viral infections [23]. Thus, ligands for TLR3 are presumably rare during bacterial infections. Nonetheless, a study by Aubry et al. [13] reported that peritoneal macrophages lacking TLR3 produced significantly less IFN-β than wild-type cells when infected with *L. monocytogenes*. The nature of the *L. monocytogenes* ligand(s) recognized by TLR3 in this setting is unclear. One possibility is that TLR3 is activated due to an association with TLR2 [13]. TLR2 recognizes lipoproteins/lipopeptides commonly found in the peptidoglycan and lipoteichoic acid of bacterial cell walls and appears to be important for recognition of *L. monocytogenes* during in vivo infections, since mice lacking TLR2 or MyD88 show impaired resistance to *L. monocytogenes* [24–27]. Furthermore, signaling from internalized TLR2 has been shown to induce type I IFN production [28, 29]. One group reported detecting type I IFN production that was dependent on IRF1 and IRF7 (but independent of IRF3) in BMDMs stimulated with the synthetic TLR2 ligand diacylated lipopeptide Pam3CSK4 [29]. Conversely, Barbalat et al. [28] reported that stimulation of TLR2 in inflammatory monocytes induced type I IFNs in response to viral but not bacterial components. Consistent with this latter report, IRF3 nuclear localization and IFN-β production were not reduced in *Tlr2*$^{-/-}$ BMMs infected with *L. monocytogenes* [11]. The lack of a role for TLR2 in type I IFN production by *L. monocytogenes*-infected BMDMs was confirmed in the study by Aubry et al. [13]. Yet, these authors also reported that TLR2 deficiency significantly reduced type I IFN production by *L. monocytogenes* infected peritoneal macrophages. Resident peritoneal macrophages are more bactericidal than BMDMs. Thus, these studies suggest TLR2 signaling may augment type I IFN production by cell types that are capable of delaying phagosomal escape of and/or digesting phagocytosed *L. monocytogenes*. Consistent with a requirement for bacterial internalization, peritoneal macrophages pre-treated with Cytochalasin D to inhibit actin mobilization before *L. monocytogenes* infection produced very little type I IFNs [13]. However, preventing internalization of *L. monocytogenes* also prevents bacterial access to the host cell cytosol and subsequent replication and stimulation of cytosolic PRRs.

Evidence for Involvement of Cytosolic PRRs

In addition to cell surface and vacuolar TLRs, macrophages and other cells can recognize microbial products using cytosolic PRRs. Recognition of microbes by different PRRs may also elicit distinct cellular responses. In the context of *L. monocytogenes* infection, it was demonstrated that a gene expression profile observed during the "early phase" (1–2 h) of BMDMs infection by virulent wild-type *L. monocytogenes* strains was also seen upon treatment of the cells with killed bacteria or Δ*hly L. monocytogenes* mutants unable to escape from vacuole compartments into the host cell cytosol [14, 30]. Several upregulated "early phase" genes (for example *Il1b*, *Tnfa*, and several chemokines) are known to be induced by TLR and NF-κB signaling pathways [14, 30], and were no longer or not as strongly induced upon infection of *MyD88*$^{-/-}$ macrophages [14]. These findings are consistent with involvement of TLR mediated pathways in the "early phase" of the macrophage response to killed or live bacteria. A distinct, "late-phase," response was also observed at 4–8 h after the infection with wild-type bacteria [14, 30]. However, this "late phase" gene expression profile was not observed after infection by killed or Δ*hly* *L. monocytogenes* strains [14, 30]. Hence, the late response appears to be indicative of infections where bacteria can access the cytosol and replicate within the macrophages. The "late phase" BMDMs genes included *Ifnb,* multiple subtypes of *Ifna*, and several additional IFN dependent genes [14, 30], and was almost entirely dependent on IRF3 activation [14]. These findings support the notion that the type I IFN response is elicited by cytosolic PRRs that are stimulated upon escape of phagocytosed wild-type *L. monocytogenes* from vacuolar compartments.

Nucleotide-Binding Oligomerization Domain-Containing (NOD) Proteins May Augment Type I IFN Responses to *L. monocytogenes*

The nucleotide-binding domain, LRR protein family referred to as NLRs includes several cytosolic and nuclear proteins. The NLR protein family has three distinct domain structures; a caspase recruitment domain (CARD) thought to regulate homotypic and heterotypic binding; a nucleotide binding domain (NBD) thought to be involved to self-oligomerization; and the LRR domain that is also thought to function in ligand binding [31]. Some LRR proteins have been shown to act as innate sensors in the detection of microbial products. For example, nucleotide-binding oligomerization domain-containing protein (NOD) 1 and NOD2 detect distinct muropeptide fragments derived from the cell wall of Gram-positive and/or Gram-negative bacteria [32]. Recognition of these fragments by NOD1 and NOD2 activates a serine/threonine kinase receptor interacting protein (RIP) 2 that is required for initiating downstream signaling and activation of NF-κB [33]. The *L. monocytogenes* cell wall contains moieties that are capable of recognition by both NOD1 and NOD2, and infection of BMDMs with *L. monocytogenes* elicits RIP2-dependent production of multiple pro-inflammatory cytokines [33–35]. However, deficiencies in NOD1, NOD2, or RIP2 do not completely ablate the cytokine response to *L. monocytogenes* indicating that this is not an essential recognition pathway [33]. Moreover, studies with RIP2 null or NOD2 null BMMs failed to reveal an essential role for these factors in mediating type I IFNs synthesis in response to *L. monocytogenes* [11, 12]. Thus, these NOD proteins do not appear to be essential for the type I IFN response elicited by replicating cytosolic *L. monocytogenes*.

However, there is some evidence that NOD proteins may, like TLRs, augment type I IFN production by *L. monocytogenes*-infected BMDMs. Specifically, while stimulation of BMDMs with synthetic MDP (the agonist for NOD2) alone elicited very little IFN-β production MDP treatment did increase IFN-β production in BMDMs transfected with *L. monocytogenes* genomic DNA by approximately twofold. The IFN-β produced in response to the DNA required expression of TBK1 and the enhancement by MDP required RIP2 [14]. To further evaluate the necessity for NOD2 in this response, BMMs were first tolerized by treatment with the TLR2 agonist, Pam3CSK4, then infected [14]. At 4 h post infection, tolerized NOD2-deficient BMMs had a twofold reduction in IFN-β synthesis compared to tolerized wild-type BMMs [14]. These findings suggest that NFκB signaling downstream of RIP2 enhances type I IFN production in *L. monocytogenes*-infected BMDMs.

Possible Contributions of RNA Helicases to the *L. monocytogenes*-Induced Type I IFN Response

During viral infections two cytosolic RNA helicases, retinoic acid inducible gene 1 (RIG-I) and melanoma differentiation-associated gene 5 (MDA5), detect viral "patterned" RNA to initiate the interferon response [36]. Both RIG-I and MDA5 contain two CARD domains required for dimerization and adaptor protein association, plus a DExD/H-box RNA helicase domain that allow for dsRNA recognition [36, 37]. Once dsRNA is detected, RIG-I or MDA5 molecules dimerize and are recruited to the mitochondria where they encounter their adaptor protein, mitochondrial antiviral signaling (MAVS) [37]. MAVS links RIG-I and MDA5 signaling to TBK1, IRF3 phosphorylation, and IFN-β synthesis [37, 38]. RIG-I is required for the type I IFN response to several ssRNA viruses while MDA5 is required for detection of another viral group, usually involving longer pieces of dsRNA [36]. Additionally, RIG-I is able to induce IFN-β production in response to cytosolic DNA when it is transcribed into a dsRNA species within the cytosol by RNA polymerase III [39].

Soon after MAVS was found to be important for viral detection, investigators asked if this adapter protein might also be involved in the type I IFN response to cytosolic *L. monocytogenes*. Studies with BMDMs from knockout mice showed that MAVS was not required to produce wild-type amounts of IFN-β in response to *L. monocytogenes* [38]. Similar conclusions were reached in studies using siRNA knockdown of MAVS in the RAW 264.7 macrophage-like cell line [40]. These findings thus argued against an essential role for MDA5 or RIG-I in the interferon response to *L. monocytogenes*. However, when Abdullah et al. [41] more directly evaluated the effects of RIG-I and MDA5 during *L. monocytogenes* challenges they found that both reacted to cytosolic *L. monocytogenes*. They reported that IFN-β production was significantly reduced in the *RigI*$^{-/-}$ BMMs and modestly reduced in *Mda5*$^{-/-}$ BMMs. However, RIG-I deletion did not completely ablate IFN-β production [41]. Additional evidence suggested that *L. monocytogenes* may actively secrete

RNA [41]. Such secreted RNA (seRNA) may also interact with RIG-I differently than RNA isolated from *L. monocytogenes* lysates [41], as seRNA induced a stronger IFN-β response when transfected into macrophages [41]. Along with secreting RNA, *L. monocytogenes* was also reported to secrete DNA, which enhanced IFN-β production through an RNA polymerase III and RIG-I dependent mechanism. These studies also included experiments using a *L. monocytogenes* SecA2 mutant (ΔSecA2) *L. monocytogenes* strain. SecA2 is a key component of an auxiliary secretory system originally identified as a protein secretion system that contributes to bacterial pathogenesis [42]. Mutants lacking SecA2 still access the cytosol of infected BMMs but do not induce the same level of IFN-β production as wild-type *L. monocytogenes*, thus the authors concluded that the SecA2 secretion system may contribute to release of nucleotides involved in activating RNA helicase pathways [41]. However, the original studies with SecA2 showed that deficiency alters bacterial morphology, impairs bacterial cell–cell spread, and impairs secretion of several *L. monocytogenes* proteins, some with demonstrated roles in pathogenicity. Thus, it is possible that one or more of these other factors contributed to the observed reduction in type I IFNs. Recent work by Hagmann et al. [43] suggests that RIG-I may play a larger role in activating type I IFN production in non-immune cell types, but additional work is needed to confirm this.

Cytosolic DNA Sensors in the Interferon Response to *L. monocytogenes* Infection

Stetson and Medzhitov [44] were first to show that IFN-β production could be induced in BMMs by a DNAse-sensitive component of *L. monocytogenes* lysates. Upon further analysis, this recognition was independent of CpG motifs in the DNA that are required for TLR9 stimulation as well as MyD88 and RIP2 [44]. Rather, the response required the sugar-phosphate DNA backbone and IRF3. These results suggested the existence of a receptor capable of sequence-independent recognition of *L. monocytogenes* DNA. This spurred a hunt for cytosolic DNA sensors that activate TBK1/IRF3 to trigger type I IFN production.

DNA-dependent activator of IFN-regulatory factors (DAI) was discovered as a potential DNA sensor through a screen for IFN inducible genes that also contained DNA binding domains [45, 46]. DAI is localized to the cytoplasm and when overexpressed in cell lines can enhance type I IFN responses to DNA. Conversely, knockdown of DAI using RNAi inhibits IFN-β induction by DNA [45]. DAI was shown to directly bind dsDNA and promote association of TBK1 and IRF3 [45]. However, siRNA knockdown of DAI had no effect on IFN-β production by human cell lines infected with *L. monocytogenes* [47]. These findings argue that DAI is not essential for the type I IFN response to *L. monocytogenes* infection, though additional studies are needed to fully understand the role this protein plays in innate DNA sensing.

LRRFIP1 is a LRR domain containing protein originally discovered for its interaction with the mammalian homolog of the gelsolin family member, *Drosophila flightless I* [48, 49]. LRRFIP1 is localized to the cytoplasm of most cells and is also known to bind dsRNA and G-C rich dsDNA [48–50]. LRRFIP1 was identified in a screen for siRNAs that reduced IFN-β production by *L. monocytogenes* infected primary peritoneal macrophages. Knockdown of LRRFIP1 reduced IFN-β secretion from infected mouse peritoneal cells by greater than 50 %, while stable knockdown in RAW 264.7 cells suppressed *L. monocytogenes* induced *Ifnb* transcripts by almost 80 % [49]. LRRFIP1 appears to act as a co-stimulator of *Ifnb* transcription. The protein was shown to interact with β-catenin to enhance its ability to bind IRF3 and recruit p300 for acetylation of histones at the *Ifnb* promoter [49]. Type I IFN production in response to *L. monocytogenes* infection was also shown to be significantly reduced in primary peritoneal macrophages deficient for β-catenin [49]. These data suggest a mechanism by which *L. monocytogenes* nucleic acids can activate LRRFIP1 to enhance *Ifnb* transcription. However, depletion of both LRRFIP1 and β-catenin failed to completely impair the type I IFN response [49].

Absent in melanoma 2 (AIM2) is another cytosolic DNA sensor. DNA binding to AIM2 causes formation of a complex called the AIM2 inflammasome, which activates caspase 1 to cleave and activate inflammatory cytokines including IL-18 and IL-1β. *L. monocytogenes* infection activates the AIM2 inflammasome, but AIM2 stimulation has not been shown to impact production of type I interferons [51–53]. In contrast, the IFI16 protein both interacts with cytosolic viral DNA and regulates production of IFN-β in both macrophages and MEFs [54, 55]. Binding of *L. monocytogenes* DNA to IFI16 has not been shown to occur, nor is it yet published whether IFI16 impacts type I responses during *L. monocytogenes* infection.

STING-Dependent Sensing of DNA or Cyclic Dinucleotides Regulates the Interferon Response to *L. monocytogenes* Infection

Stimulator of interferon genes (STING), also called MITA, MPYS, or ERIS, is an evolutionarily conserved protein that contains five transmembrane regions and is localized in the endoplasmic reticulum [56–59]. The involvement of STING in type I IFN responses was first discovered in a screen where full length cDNA expression vectors were transfected into 293T cells containing a luciferase construct driven by the IFN-β promoter [56, 57]. Over-expression of STING increased IRF3 activation and IFN-β production in response to viral challenges [56, 57, 59]. RNAi knockdown or a direct knockout of STING resulted in a decreased activation of IRF3 and decreased IFN-β production, ultimately leading to increased viral susceptibility [56, 57, 59]. In fact, STING expression levels correlated with the degree of inhibited viral replication [57]. Upon viral infection, STING dimerizes and directly interacts with TBK1 in immunoprecipitation experiments [56, 57, 59]. STING also enhances interaction of TBK1 and IRF3 and both of these factors are required for STING-induced type I IFN production [56, 57, 59]. To identify stimuli leading to STING-dependent induction of IFN-β, MEFs derived from wild-type and STING$^{-/-}$ mice were transfected with various DNA ligands. STING expression enhanced IFN-β synthesis in response to cytosolic delivery of both viral and bacterial DNA, as well as synthetic non-CpG dsDNA, but not dsRNA [60]. Macrophages and dendritic cells isolated from *Sting*$^{-/-}$ mice also demonstrated significantly reduced or undetectable levels of IFN-I when transfected with synthetic DNA or infected with *L. monocytogenes* [60–62].

STING does not appear to be a direct sensor of DNA. Rather, cyclic dinucleotides—which act as second messengers in a number of bacterial species—are able to induce type I IFN production in a STING-dependent manner [61–63]. STING binds radiolabeled cyclic diguanylate monophosphate (c-di-GMP) in a manner competed by unlabeled cyclic dinucleotides but not other nucleic acids such as dsDNA [63]. Another study found that biotinylated c-di-GMP and c-di-AMP also bound to the DEAD-box helicase, DDX41, with a higher affinity than to STING [64]. Unlike STING, DDX41 also bound cytosolic DNA. Mouse or human cells deficient for DDX41 also showed decreased IFN-β responses to *L. monocytogenes* infection or cytosolic delivery of c-di-AMP and c-di-GMP [64]. Yet, STING was still required for type I interferon production to these stimuli as well as synthetic dsDNA and DNA viruses [65]. Since DDX41 also binds to STING, it may act as a co-factor to regulate STING-dependent type I IFN responses [64, 65].

Evidence suggests that cyclic di-nucleotides are actively released from replicating *L. monocytogenes* [66]. The release of c-di-AMP from *L. monocytogenes* appears to be mediated by a family of multidrug efflux transporters (MDRs) [66]. *L. monocytogenes* strains containing increased or reduced expression of MDRs such as MdrM show corresponding increases and reductions in their ability to elicit IFN-β production by infected BMDMs [66]. *L. monocytogenes* production of c-di-AMP requires a diadenylate cyclase (DacA), which is required for establishment and optimal growth within mammalian cells, as well as the overall stability of its bacterial cell wall [67]. Strains deficient in DacA are significantly attenuated during infections of mice, yet still induce type I IFN production [67]. The residual activation of type I IFNs could reflect the release of c-di-GMP other cyclic di-nucleotides that activate STING, or the release of DNA or RNA. Knockdown of STING in RAW 264.7 cells and BMMs derived from a *Sting*$^{-/-}$ mouse significantly decreased IRF3 activation and IFN-β production in response to *L. monocytogenes* infection or cytosolic delivery of c-di-AMP and c-di-GMP [61]. During systemic *L. monocytogenes* infection in mice, STING deficiency also impacted early production of type I interferons as *Sting*$^{-/-}$ mice had significantly reduced IFN-β in the sera 8 h post infection [61]. Similar results were independently observed using an *N*-Ethyl-*N-Nitrosourea* (ENU) generated mouse with a loss of function mutation in STING [62]. These data indicate the importance of STING in the initial type I interferon response to cytosolic *L. monocytogenes* and suggest this could be due to bacterial release of cyclic di-nucleotides. It is also possible that bacterial DNA released into the cytosol could contribute to this STING-dependent response. It was recently shown that cytosolic or viral DNA can be processed into a "non-canonical" 2′–5′ linked cyclic dinucleotide, cyclic guanosine monophosphate–adenosine monophosphate (cGAMP) by an enzyme named cGAMP synthase (cGAS) [68, 69]. This contrasts with the canonical 3′–5′ linkage seen in the cyclic di-nucleotides produced by bacteria. Like bacterial cyclic di-nucleotides, cGAMP binds STING and does so in a manner competed by high concentrations of unlabeled c-di-GMP, c-di-AMP, and cGAMP, but not by DNA [68]. Furthermore, over-expression of cGAS induced IFN-β production that was dependent on STING expression and knockdown of cGAS significantly reduced IRF3 activation and *Ifnb* transcription in response to DNA [69]. Whether cGAS might also play a role in the type I interferon response by macrophages or other cell types infected with *L. monocytogenes* is not yet known.

Biological Consequences of Type I IFN Production

Type I interferons bind a common cell surface receptor to alter gene expression in a manner that induces an antiviral state that increases cell intrinsic resistance to viral replication. Thus, production and response to these interferons increases host resistance to numerous viral infections. The opposite occurs during infections by *L. monocytogenes* and several other bacteria, where responsiveness to type I IFNs is actually detrimental to the host [9, 70–72]. Mice deficient in IFNAR and IRF3 are also significantly more resistant to *L. monocytogenes* challenge [9, 70]. In wild-type mice, treatment with the type I interferon-inducing synthetic dsRNA agonist poly:IC also significantly increased *L. monocytogenes* titres in both the livers and spleens [9]. These results indicate that type I IFN production and responsiveness exacerbate *L. monocytogenes* pathogenicity. However, although STING-deficient mice have reduced production of IFN-β early after *L. monocytogenes* infection, they were not more resistant to *L. monocytogenes* and showed similar bacterial burdens in the both the livers and spleens compared to STING sufficient mice [61]. These results suggest that the lack of IFN production very early after infection is not sufficient to increase host resistance and also that redundancy exists in the pathways required for *L. monocytogenes* induced type I IFN during systemic infection.

Multiple mechanisms have been proposed to account for the deleterious effects of type I IFNs during *L. monocytogenes* challenge. O'Connell et al. [9] observed that type I IFN signaling increased the expression of several pro-apoptotic genes such as TRAIL, PML, and Daxx. Additionally, more macrophages and inflammatory monocytes were found in the spleens of *L. monocytogenes* infected *Ifnar*$^{-/-}$ mice compared to wild type [9]. These results suggested to the authors that type I IFNs may be deleterious because they induce apoptosis of monocytes within the spleens. Another group observed decreased terminal deoxynucleotidyl transferase-mediated dUTP nick-end labeling (TUNEL) staining in the spleens of *Ifnar*$^{-/-}$ mice after 2–3 days of *L. monocytogenes* infection [73]. These authors concluded that the apoptotic cells were lymphocytes and not monocytes, and thus that IFN-induced apoptosis of lymphocytes was deleterious to the host [73]. In contrast, Auerbuch et al. [70] reported increased numbers of splenic CD11b+ cells secreting the pro-inflammatory cytokine, TNFα, within the spleens of *L. monocytogenes* infected *Ifnar*$^{-/-}$ mice, leading them to suggest type I IFN signaling suppresses accumulation of TNFα producing monocytes that might protect against *L. monocytogenes* infection [70].

In contrast to type I IFNs, the type II IFN or IFNγ is critical for the pro-inflammatory activation of macrophages. IFNγ enhances macrophage ability to kill bacteria, increases their secretion of pro-inflammatory cytokines such as TNFα and IL-12, and increases expression of MHC class II and co-stimulatory molecules [74]. IFNγ signals through a heterodimeric receptor IFN gamma receptor (IFNGR). During a *L. monocytogenes* infection, it was observed that the IFNGR was selectively down regulated from the surface of myeloid cells, but not T cells [71, 75]. This phenomenon was observed both in vivo and in vitro upon challenge with *L. monocytogenes* and was mediated by type I IFNs [71, 75]. BMDMs derived from wild-type mice also decrease surface expression of IFNGR upon stimulation with IFN-β [71, 75]. The suppression of the IFNGR receptor decreased the responsiveness of the myeloid cells to IFNγ, potentially suppressing pro-inflammatory activation of macrophages and decreasing their ability to clear bacterial infections [71]. This thus represents an additional potential mechanism to account for the ability of type I IFNs to increase host susceptibility to bacterial infections.

Mechanistically, down regulation of the IFNGR involves transcriptional silencing by type I IFNs [71, 75]. Kearney et al. [75] demonstrated that IFN-β stimulation silences new transcription at the *ifngr* locus in macrophages, as indicated by loss of activated RNA polymerase II at the transcriptional start site as well as epigenetic marks indicative of condensed chromatin. Additionally, recruitment of early growth response factor 3 (Egr3) to the *ifngr* promoter was observed shortly after IFN-β treatment [75]. Egr3 can act as a activator or repressor of transcription [76–79]. Association of Egr proteins with the NGFI-A binding protein, Nab1, causes transcriptional silencing and Nab1 was recruited to the *ifngr* promoter shortly after Egr3 [75]. Knockdown of Nab1 in mouse RAW 264.7 macrophages prevented IFNGR down regulation in response to type I IFN stimulation [75]. These data provide evidence of a direct antagonistic effect between type I and type II IFNs in myeloid cells and suggest this antagonism lowers myeloid cell responsiveness to IFNγ and thus host resistance. However, there is not yet direct evidence to support whether one of these possible mechanisms is responsible for the increased bacterial burdens in response to type I IFNs.

Conclusions

Sensing of microbial products is important for host defense against pathogens. Yet, sensing of *L. monocytogenes* and other bacterial pathogens appears to be deleterious to the host when this leads to the production of type I IFNs. *L. monocytogenes* may thus promote such sensing as there is evidence it actively secretes RNA, DNA, and cyclic di-nucleotides that are recognized by cytosolic PRRs including RIG-I, STING, DDX41, IFI16, and cGAS. STING expression is most critical for the induction of IFN-I in cultured macrophages, but whether this is through a direct interaction with *L. monocytogenes* c-di-AMP is uncertain. However, mice lacking STING still produce type I IFNs in response to *L. monocytogenes* infection, highlighting the redundancy in these pathways mediating detection of pathogen-derived molecules and triggering of IFN-β production. The creation of double and triple knockout mice would provide a valuable tool to further dissect which sensing pathways are most crucial for *L. monocytogenes* sensing in vivo. Further understanding of how type I IFNs are triggered, and the effects they have on host biology, is essential for improving our knowledge of and ability to improve host resistance to bacterial infections.

References

1. McCollum JT, Cronquist AB, Silk BJ et al (2013) Multistate outbreak of listeriosis associated with cantaloupe. N Engl J Med 369:944–953
2. Mostowy S, Cossart P (2009) Cytoskeleton rearrangements during *Listeria* infection: clathrin and septins as new players in the game. Cell Motil Cytoskeleton 66:816–823
3. Hamon M, Bierne H, Cossart P (2006) *Listeria monocytogenes*: a multifaceted model. Nat Rev Microbiol 4:423–434
4. Ireton K (2007) Entry of the bacterial pathogen *Listeria monocytogenes* into mammalian cells. Cell Microbiol 9:1365–1375
5. Schnupf P, Portnoy DA (2007) Listeriolysin O: a phagosome-specific lysin. Microbes Infect 9:1176–1187
6. Portnoy D, Jacks P, Hinrichs D (1988) Role of hemolysin for the intracellular growth of *Listeria monocytogenes*. J Exp Med 167:1459–1471
7. Berche P, Gaillard JL, Sansonetti PJ (1987) Intracellular growth of *Listeria monocytogenes* as a prerequisite for in vivo induction of T cell-mediated immunity. J Immunol 138:2266–2271
8. Taniguchi T, Ogasawara K, Takaoka A, Tanaka N (2001) IRF family of transcription factors as regulators of host defense. Annu Rev Immunol 19:623–655
9. O'Connell RM, Saha SK, Vaidya SA et al (2004) Type I interferon production enhances susceptibility to *Listeria monocytogenes* infection. J Exp Med 200:437–445
10. Garifulin O, Qi Z, Shen H et al (2007) Irf3 polymorphism alters induction of interferon beta in response to *Listeria monocytogenes* infection. PLoS Genet 3:1587–1597
11. O'Connell R, Vaidya S, Perry AK et al (2005) Immune activation of type I IFNs by *Listeria monocytogenes* occurs independently of TLR4, TLR2, and receptor interacting protein 2 but involves TANK-binding kinase. J Immunol 174:1602–1607
12. Stockinger S, Reutterer B, Schaljo B et al (2004) IFN regulatory factor 3-dependent induction of type I IFNs by intracellular bacteria is mediated by a TLR- and Nod2-independent mechanism. J Immunol 173:7416–7425
13. Aubry C, Corr SC, Wienerroither S et al (2012) Both TLR2 and TRIF contribute to interferon-β production during *Listeria* infection. PLoS One 7:e33299
14. Leber JH, Crimmins GT, Raghavan S et al (2008) Distinct TLR- and NLR-mediated transcriptional responses to an intracellular pathogen. PLoS Pathog 4:e6
15. Pomerantz JL, Baltimore D (1999) NF-kappaB activation by a signaling complex containing TRAF2, TANK and TBK1, a novel IKK-related kinase. EMBO J 18:6694–6704
16. Fitzgerald KA, McWhirter SM, Faia KL et al (2003) IKKepsilon and TBK1 are essential components of the IRF3 signaling pathway. Nat Immunol 4:491–496

17. Sharma S, tenOever BR, Grandvaux N et al (2003) Triggering the interferon antiviral response through an IKK-related pathway. Science 300:1148–1151
18. Perry AK, Chow EK, Goodnough JB et al (2004) Differential requirement for TANK-binding kinase-1 in type I interferon responses to Toll-like receptor activation and viral infection. J Exp Med 199:1651–1658
19. Yu T, Yi Y-S, Yang Y et al (2012) The pivotal role of TBK1 in inflammatory responses mediated by macrophages. Mediators Inflamm 2012:979105
20. Takeda K, Akira S (2005) Toll-like receptors in innate immunity. Int Immunol 17:1–14
21. Doyle S, Vaidya S, O'Connell R et al (2002) IRF3 mediates a TLR3/TLR4-specific antiviral gene program. Immunity 17:251–263
22. Takeuchi O, Hoshino K, Kawai T et al (1999) Differential roles of TLR2 and TLR4 in recognition of Gram-negative and Gram-positive bacterial cell wall components. Immunity 11: 443–451
23. Alexopoulou L, Holt AC, Medzhitov R, Flavell RA (2001) Recognition of double-stranded RNA and activation of NF-kappaB by Toll-like receptor 3. Nature 413:732–738
24. Seki E, Tsutsui H, Tsuji NM et al (2002) Critical roles of myeloid differentiation factor 88-dependent proinflammatory cytokine release in early phase clearance of *Listeria monocytogenes* in mice. J Immunol 169:3863–3868
25. Janot L, Secher T, Torres D et al (2008) CD14 works with toll-like receptor 2 to contribute to recognition and control of *Listeria monocytogenes* infection. J Infect Dis 198:115–124
26. Torres D, Barrier M, Bihl F et al (2004) Toll-like receptor 2 is required for optimal control of *Listeria monocytogenes* infection. Infect Immun 72:2131–2139
27. Edelson BT, Unanue ER (2002) MyD88-dependent but Toll-like receptor 2-independent innate immunity to *Listeria*: no role for either in macrophage listericidal activity. J Immunol 169: 3869–3875
28. Barbalat R, Lau L, Locksley RM, Barton GM (2009) Toll-like receptor 2 on inflammatory monocytes induces type I interferon in response to viral but not bacterial ligands. Nat Immunol 10:1200–1207
29. Dietrich N, Lienenklaus S, Weiss S, Gekara NO (2010) Murine toll-like receptor 2 activation induces type I interferon responses from endolysosomal compartments. PLoS One 5:e10250
30. McCaffrey RL, Fawcett P, O'Riordan M et al (2004) A specific gene expression program triggered by Gram-positive bacteria in the cytosol. Proc Natl Acad Sci U S A 101:11386–11391
31. Creagh EM, O'Neill LAJ (2006) TLRs, NLRs and RLRs: a trinity of pathogen sensors that co-operate in innate immunity. Trends Immunol 27:352–357
32. Kufer TA, Banks DJ, Philpott DJ (2006) Innate immune sensing of microbes by Nod proteins. Ann N Y Acad Sci 1072:19–27
33. Park J-H, Kim Y-G, McDonald C et al (2007) RICK/RIP2 mediates innate immune responses induced through Nod1 and Nod2 but not TLRs. J Immunol 178:2380–2386
34. Kobayashi K, Inohara N, Hernandez LD et al (2002) RICK/Rip2/CARDIAK mediates signalling for receptors of the innate and adaptive immune systems. Nature 416:194–199
35. Chin AI, Dempsey PW, Bruhn K et al (2002) Involvement of receptor-interacting protein 2 in innate and adaptive immune responses. Nature 416:190–194
36. Rehwinkel J, Reis e Sousa C (2010) RIGorous detection: exposing virus through RNA sensing. Science 327:284–286
37. Hiscott J, Lin R, Nakhaei P, Paz S (2006) MasterCARD: a priceless link to innate immunity. Trends Mol Med 12:53–56
38. Sun Q, Sun L, Liu H-H et al (2006) The specific and essential role of MAVS in antiviral innate immune responses. Immunity 24:633–642
39. Chiu Y-H, Macmillan JB, Chen ZJ (2009) RNA polymerase III detects cytosolic DNA and induces type I interferons through the RIG-I pathway. Cell 138:576–591
40. Soulat D, Bauch A, Stockinger S et al (2006) Cytoplasmic *Listeria monocytogenes* stimulates IFN-beta synthesis without requiring the adapter protein MAVS. FEBS Lett 580:2341–2346
41. Abdullah Z, Schlee M, Roth S et al (2012) RIG-I detects infection with live *Listeria* by sensing secreted bacterial nucleic acids. EMBO J 31:4153–4164

42. Lenz LL, Mohammadi S, Geissler A, Portnoy DA (2003) SecA2-dependent secretion of autolytic enzymes promotes *Listeria monocytogenes* pathogenesis. Proc Natl Acad Sci U S A 100:12432–12437
43. Hagmann CA, Herzner AM, Abdullah Z et al (2013) RIG-I detects triphosphorylated RNA of *Listeria monocytogenes* during infection in non-immune cells. PLoS One 8:e62872
44. Stetson DB, Medzhitov R (2006) Recognition of cytosolic DNA activates an IRF3-dependent innate immune response. Immunity 24:93–103
45. Takaoka A, Wang Z, Choi MK et al (2007) DAI (DLM-1/ZBP1) is a cytosolic DNA sensor and an activator of innate immune response. Nature 448:501–505
46. Yanai H, Savitsky D, Tamura T, Taniguchi T (2009) Regulation of the cytosolic DNA-sensing system in innate immunity: a current view. Curr Opin Immunol 21:17–22
47. Lippmann J, Rothenburg S, Deigendesch N et al (2008) IFNbeta responses induced by intracellular bacteria or cytosolic DNA in different human cells do not require ZBP1 (DLM-1/DAI). Cell Microbiol 10:2579–2588
48. Wilson SA, Brown EC, Kingsman AJ, Kingsman SM (1998) TRIP: a novel double stranded RNA binding protein which interacts with the leucine rich repeat of flightless I. Nucleic Acids Res 26:3460–3467
49. Yang P, An H, Liu X et al (2010) The cytosolic nucleic acid sensor LRRFIP1 mediates the production of type I interferon via a beta-catenin-dependent pathway. Nat Immunol 11:487–494
50. Suriano A, Sanford A, Kim N (2005) GCF2/LRRFIP1 represses tumor necrosis factor alpha expression. Mol Cell Biol 25:9073–9081
51. Hornung V, Ablasser A, Charrel-Dennis M et al (2009) AIM2 recognizes cytosolic dsDNA and forms a caspase-1-activating inflammasome with ASC. Nature 458:514–518
52. Hornung V, Latz E (2010) Critical functions of priming and lysosomal damage for NLRP3 activation. Eur J Immunol 40:620–623
53. Tsuchiya K, Hara H, Kawamura I et al (2010) Involvement of absent in melanoma 2 in inflammasome activation in macrophages infected with *Listeria monocytogenes*. J Immunol 185:1186–1195
54. Unterholzner L, Keating SE, Baran M et al (2010) IFI16 is an innate immune sensor for intracellular DNA. Nat Immunol 11:997–1004
55. Schattgen SA, Fitzgerald KA (2011) The PYHIN protein family as mediators of host defenses. Immunol Rev 243:109–118
56. Ishikawa H, Barber GN (2008) STING is an endoplasmic reticulum adaptor that facilitates innate immune signalling. Nature 455:674–678
57. Zhong B, Yang Y, Li S et al (2008) The adaptor protein MITA links virus-sensing receptors to IRF3 transcription factor activation. Immunity 29:538–550
58. Jin L, Waterman PM, Jonscher KR et al (2008) MPYS, a novel membrane tetraspanner, is associated with major histocompatibility complex class II and mediates transduction of apoptotic signals. Mol Cell Biol 28:5014–5026
59. Sun W, Li Y, Chen L et al (2009) ERIS, an endoplasmic reticulum IFN stimulator, activates innate immune signaling through dimerization. Proc Natl Acad Sci U S A 106:8653–8658
60. Ishikawa H, Ma Z, Barber GN (2009) STING regulates intracellular DNA-mediated, type I interferon-dependent innate immunity. Nature 461:788–792
61. Jin L, Hill KK, Filak H et al (2011) MPYS is required for IFN response factor 3 activation and type I IFN production in the response of cultured phagocytes to bacterial second messengers cyclic-di-AMP and cyclic-di-GMP. J Immunol 187:2595–2601
62. Sauer J-D, Sotelo-Troha K, von Moltke J et al (2011) The *N*-ethyl-*N*-nitrosourea-induced Goldenticket mouse mutant reveals an essential function of Sting in the in vivo interferon response to *Listeria monocytogenes* and cyclic dinucleotides. Infect Immun 79:688–694
63. Burdette DL, Monroe KM, Sotelo-Troha K et al (2011) STING is a direct innate immune sensor of cyclic di-GMP. Nature 478:515–518
64. Parvatiyar K, Zhang Z, Teles RM et al (2012) The helicase DDX41 recognizes the bacterial secondary messengers cyclic di-GMP and cyclic di-AMP to activate a type I interferon immune response. Nat Immunol 13:1155–1161

65. Zhang Z, Yuan B, Bao M et al (2011) The helicase DDX41 senses intracellular DNA mediated by the adaptor STING in dendritic cells. Nat Immunol 12:959–965
66. Woodward JJ, Iavarone AT, Portnoy DA (2010) c-di-AMP secreted by intracellular *Listeria monocytogenes* activates a host type I interferon response. Science 328:1703–1705
67. Witte CE, Whiteley AT, Burke TP et al (2013) Cyclic di-AMP is critical for *Listeria monocytogenes* growth, cell wall homeostasis, and establishment of infection. MBio 4:e00282-13
68. Wu J, Sun L, Chen X et al (2013) Cyclic GMP-AMP is an endogenous second messenger in innate immune signaling by cytosolic DNA. Science 339:826–830
69. Sun L, Wu J, Du F et al (2013) Cyclic GMP-AMP synthase is a cytosolic DNA sensor that activates the type I interferon pathway. Science 339:786–791
70. Auerbuch V, Brockstedt DG, Meyer-Morse N et al (2004) Mice lacking the type I interferon receptor are resistant to *Listeria monocytogenes*. J Exp Med 200:527–533
71. Rayamajhi M, Humann J, Penheiter K et al (2010) Induction of IFN-alpha/beta enables *Listeria monocytogenes* to suppress macrophage activation by IFN-gamma. J Exp Med 207: 327–337
72. Rayamajhi M, Humann J, Kearney S et al (2010) Antagonistic crosstalk between type I and II interferons and increased host susceptibility to bacterial infections. Virulence 1:418–422
73. Carrero JA, Calderon B, Unanue ER (2004) Type I interferon sensitizes lymphocytes to apoptosis and reduces resistance to Listeria infection. J Exp Med 200:535–540
74. Mosser DM, Edwards JP (2008) Exploring the full spectrum of macrophage activation. Nat Rev Immunol 8:958–969
75. Kearney SJ, Delgado C, Eshleman EM et al (2013) Type I IFNs downregulate myeloid cell IFN-γ receptor by inducing recruitment of an early growth response 3/NGFI-A binding protein 1 complex that silences *ifngr1* transcription. J Immunol 191:3384–3392
76. Swirnoff A, Milbrandt J (1995) DNA-binding specificity of NGFI-A and related zinc finger transcription factors. Mol Cell Biol 15:2275–2287
77. O'Donovan KJ, Tourtellotte WG, Millbrandt J, Baraban JM (1999) The EGR family of transcription-regulatory factors: progress at the interface of molecular and systems neuroscience. Trends Neurosci 22:167–173
78. Yu J, de Belle I, Liang H, Adamson ED (2004) Coactivating factors p300 and CBP are transcriptionally crossregulated by Egr1 in prostate cells, leading to divergent responses. Mol Cell 15:83–94
79. Huang RP, Fan Y, deBelle I et al (1998) Egr-1 inhibits apoptosis during the UV response: correlation of cell survival with Egr-1 phosphorylation. Cell Death Differ 5:96–106

Innate Immune and Type I IFN Responses During *Legionella pneumophila* Infection

Jan Naujoks and Bastian Opitz

Introduction

Legionella pneumophila can cause a flu-like disease, called Pontiac fever, and a severe pneumonia known as Legionnaires' disease or legionellosis. Legionnaires' disease can affect ambulatory and hospitalized patients and is associated with high mortality rates ranging from 10 to 38 % [1, 2]. Risk factors associated with the occurrence of this disease include old age, solid organ transplantation, smoking, a history of cancer or hematologic malignancies, steroid therapy, other immunosuppressive treatments, and diabetes mellitus [1]. The numbers of patients with those risk factors as well as the number of reported cases of legionellosis are increasing [3, 4].

L. pneumophila is a facultative intracellular bacterium and a parasite of amoeba allowing the bacteria to persist in the environment [5, 6]. Infection develops following inhalation of *L. pneumophila*-contaminated aerosols into the human lung, phagocytic uptake, and intracellular growth of the bacteria in alveolar macrophages. In the phagosome, *L. pneumophila* circumvents the phagolysosomal pathway and recruits ER vesicles and mitochondria to the vacuole, forming the *Legionella-containing* vacuole (LCV). The formation of LCV is dependent on *Legionella* effector proteins that are translocated into the host cytoplasm by a type IV secretion system (T4SS) encoded by *dot/icm* genes [5, 6]. Evading the antibacterial lysosomal activity, the bacterium is able to replicate in the LCV.

While intracellular pathogens, such as *L. pneumophila,* manipulate host cell processes in order to establish an intracellular niche for their survival and replication, the host has evolved defense mechanisms that restrict the infection. Among them, intracellular defense pathways appear to play a major role to fight *L. pneumophila* pneumonia (see next paragraph). The balance between bacterial virulence strategies

J. Naujoks • B. Opitz (✉)
Department of Internal Medicine/Infectious Diseases and Pulmonary Medicine,
Charité Universitätsmedizin Berlin, Augustenburger Platz 1, 13353 Berlin, Germany
e-mail: bastian.opitz@charite.de

© Springer International Publishing Switzerland 2014

D. Parker (ed.), *Bacterial Activation of Type I Interferons*,
DOI 10.1007/978-3-319-09498-4_3

and defense pathways of the host determines the outcome of such bacterial encounters, resulting in microbial clearance, the mild Pontiac fever or establishment of Legionnaires' disease.

Innate Immunity to *L. pneumophila* Infection

L. pneumophila is recognized by several transmembrane and cytosolic pattern recognition receptors that cooperatively mediate protective immune responses [7]. The transmembrane Toll-like receptors (TLRs) -2, -5, and -9 detect bacterial cell-wall components, flagellin and unmethylated CpG-rich DNA, respectively [8–12]. TLRs stimulate production of several NF-κB-dependent cytokines such as TNFα which contributes to resistance of mice towards *L. pneumophila* infection [13–16]. Several studies demonstrated that mice deficient in TLR2 and the other above-mentioned TLRs alone or in different combinations have defects in the defense against *L. pneumophila* compared to wild-type mice [9, 10, 15–18]. The NOD-like receptors (NLRs) NOD1 and NOD2 are activated by *Legionella* peptidoglycan that might get access to the cytosol through the T4SS. Mice deficient in both NODs or in the shared signaling mediator RIP2 show impaired neutrophil recruitment and attenuated bacterial clearance during pneumonia [19, 20].

Other cytosolic sensors of *L. pneumophila* in macrophages are the canonical NAIP5 and NLRP3 inflammasomes and the non-canonical caspase-11-dependent inflammasome. Regarding the first mentioned inflammasome, different alleles of NAIP5 have long been known to determine whether a mouse is resistant or (moderately) susceptible to *Legionella* infection, underlying the importance of this NLR in the host defense mechanism [21, 22]. NAIP5 forms together with the NLR molecule NLRC4 the NAIP5 inflammasome, which can additionally contain the adapter molecule ASC and caspase-1 [23–27]. This multi-protein complex is activated by the flagellin delivered by the T4SS. The activation contributes to production of IL-1β and IL-18 and leads to the growth restriction of wild-type (but not flagellin-deficient) *L. pneumophila* in macrophages of most mouse strains (e.g. C57Bl/6). The growth restriction is dependent on the caspase-1-mediated cell death called pyroptosis and on the enhancement of the fusion of LCVs with lysosomes [24, 28–31]. The canonical NLRP3 inflammasome, consisting of NLRP3, ASC, and caspase-1, is also activated by *L. pneumophila* and controls IL-1β and IL-18 production, but its function in controlling infection in vivo appears to be less essential [32, 33].

Furthermore, *L. pneumophila* stimulates a cytosolic non-canonical caspase-11-dependent inflammasome depending on its T4SS [32, 34, 35]. The exact mode of action of this inflammasome and its molecular components are unclear. Upon *L. pneumophila* infection of macrophages, the caspase-11 inflammasome contributes to the NLRP3 inflammasome-mediated IL-1β production and cell death, and stimulates a NLRP3-independent cell death and IL-1α release [32, 34, 35]. Moreover, caspase-11 has been indicated to stimulate fusion of LCVs with lysosomes [36]. Thus, different inflammasomes contribute to the macrophage-intrinsic defense against *L. pneumophila*. The innate immune response to *L. pneumophila* is further shaped by translational inhibition and biasing to favor production of some pro-inflammatory mediators [37, 38]. This translational regulation is dependent on the T4SS and possibly on some effector proteins and/or on effector protein-independent inhibition of the mTOR pathway.

The function of neutrophils in *Legionella* infection is incompletely understood. Recruitment of neutrophils to the lung during infection is dependent on TLR- and NOD1/2-dependent chemokine production [19], release of IL-1α (in addition to IL-1β) by hematopoietic cells [39], activation of IL-1 receptor (IL-1R) and production of chemokines by non-hematopoietic cells [40]. *Legionella* spp. might be resistant to neutrophilic killing [41, 42], but antibody-mediated depletion of neutrophils impairs clearance of *L. pneumophila* from the lung at later time points [43]. This might be related to the production of cytokines such as IL-18 by neutrophils which together with IL-12 activates natural killer (NK) cells to produce the host protective type II IFN (IFNγ; [44, 45]).

IFN-γ activates macrophages to restrict *L. pneumophila* replication. Mice lacking IFN-γ or its receptor IFNGR are susceptible towards *L. pneumophila* infection [46–48]. The IFN-γ-mediated defense in macrophages most likely depends on STAT1 homodimerization, binding to gamma IFN-activated sites (GAS) in gene promoters and up-regulation of IFN-stimulated antibacterial genes (ISGs). Yet, the identity of these antibacterial factors and their modes of action in *L. pneumophila* infection are still unknown.

Besides neutrophils and NK cells, plasmacytoid dendritic cells (pDCs) have been demonstrated to have an important contribution to the restriction of *L. pneumophila* in mice [49]. These cells were recruited during infection and depletion of pDCs significantly decreases bacterial clearance from the lung. Interestingly, even though pDCs are well known for their ability to produce type I IFNs upon viral infection, the protective effect of pDCs on *L. pneumophila* infection is independent of these cytokines [49].

Production of Type I IFNs in *L. pneumophila* Infection

In addition to the above-mentioned pathways, *L. pneumophila* infection of macrophages is also detected by cytosolic nucleic acid sensors, and restricted by subsequently produced type I IFNs. We and others have previously shown that host cells infected with *L. pneumophila* produce type I IFNs [50, 51]. This response requires bacterial uptake and expression of the bacterial T4SS, but is independent of bacterial replication and the IcmS-dependently translocated bacterial proteins [50–52]. Although the sensor molecule is still unknown, several lines of evidence suggest that bacterial DNA is the molecule that is detected into the host cell cytosol and triggers type I IFN production. First, intracellular delivery of purified *Legionella* DNA into macrophages stimulates a similar type I IFN production as infection with viable bacteria [50, 52]. Second, type I IFN responses to *L. pneumophila* are dependent on the T4SS, and the T4SS has been shown to transfer DNA to recipient bacteria [53]. Third, digestion of *Legionella* extracts with DNAse (but not RNAse or proteinase) inhibits their ability to induce IFN-β expression [50, 52]. Fourth, the expression of the T4SS effector molecule SdhA negatively correlates with both, *Legionella* DNA release into the host cell cytosol and type I IFN responses [54–56]. Fifth, *L. pneumophila*-induced type I IFN production is significantly reduced in macrophages after gene-silencing of STING [52]. STING (also called MPYS, ERIS, and MITA) is an ER-anchored molecule that serves as a key adapter protein for most cytosolic DNA sensing pathways [57, 58]. These cytosolic DNA sensors include cyclic-AMP-GMP synthase (cGAS), DAI, IFI16, DDX41, and RNA polymerase III/RIG-I [59–64]. While DAI is not involved [65], and the function of RNA polymerase III/RIG-I is controversial [55, 64] the role of the other DNA sensors in *L. pneumophila*-induced type I IFN responses needs to be examined.

In addition to DNA sensing, detection of other bacterial molecules might contribute to stimulation of type I IFN responses during *L. pneumophila* infection. For example, *Legionella* second messenger molecules such as the cyclic dinucleotide c-di-GMP might also be involved in triggering STING-dependent innate immune responses, as STING also serves as receptor for cyclic di-GMP (c-di-GMP) and c-di-AMP [66, 67]. Indeed, a recent study found that the amount of IFN-β expression in macrophages positively correlated with c-di-GMP levels in *L. pneumophila* [68]. Moreover, one study indicated that recognition of *Legionella* RNA by the cytosolic RNA receptors RIG-I and MDA5 stimulated type I IFN response in macrophages [55]. This is all summarized in Fig. 1.

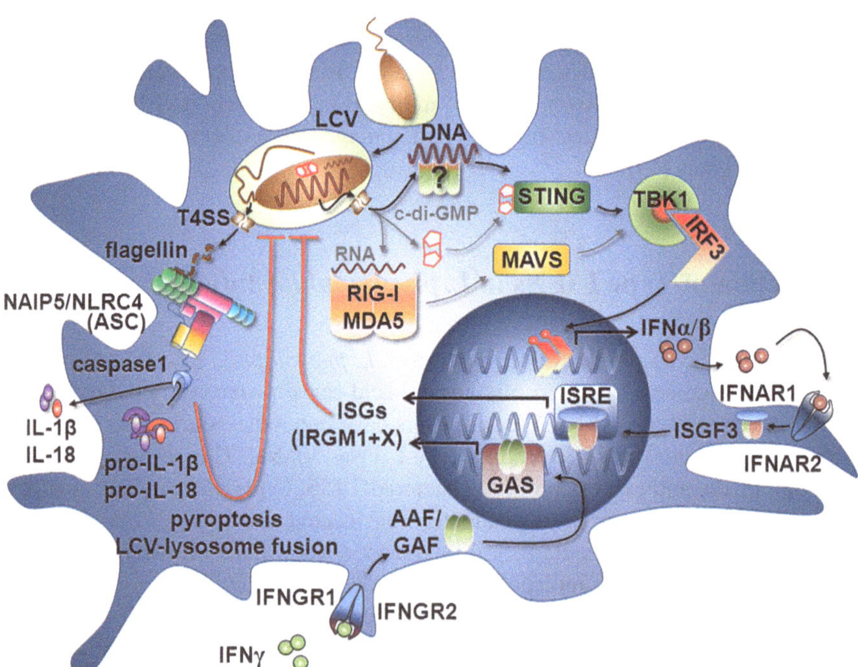

Fig. 1 Overview of the type I IFN pathway and other important mechanisms that restrict *L. pneumophila* in macrophage as discussed in the main text

Thus, further work is needed to fully understand which bacterial molecules and which host cell receptors stimulate type I IFN production during *L. pneumophila* infections. Downstream of the sensor(s) and adapter molecules, the kinase TBK1 and the transcription factor IRF3 are required for stimulating type I IFN responses to *L. pneumophila* [50, 51].

Function of Type I IFNs in *L. pneumophila* Infection

The first evidence for a host-protective function of type I IFNs in *L. pneumophila* infections came from a study by Schiavoni et al. [69]. The authors showed that treatment with IFN-β inhibited growth of *L. pneumophila* in permissive murine A/J macrophages, whereas addition of blocking anti-IFNα/β antibodies allowed bacterial growth in nonpermissive cells [69]. Subsequently, we showed that inhibition of IRF3 expression by RNAi, and thus suppression of type I IFN production, resulted in enhanced *L. pneumophila* replication in human cells. The enhanced bacterial replication in IRF3-depleted cells could be reversed by treatment of the cells with exogenous IFN-β [51]. Similarly, mouse macrophages deficient in IRF3 or IFNAR (in C57Bl/6 background) allowed bacterial replication, whereas wild-type macrophages inhibit *L. pneumophila* replication [52, 70, 71]. These studies together show that endogenously produced type I IFN acts in an autocrine fashion to activate a macrophage-intrinsic antibacterial defense pathway that limits *L. pneumophila* infection. Importantly, recombinant IFN-β inhibits the growth of flagellin-deficient *Legionella* in wild-type macrophages, indicating that the type I IFN-mediated antibacterial defense acts independently of the flagellin-detecting NAIP5 inflammasome [52]. Whereas activity of NAIP5 and type I IFN pathways efficiently suppresses bacterial replication, functional defects in one of those pathways lead to a substantial growth of *L. pneumophila* in macrophages.

The mechanism of the type I IFN-mediated resistance pathway is still incompletely understood but appears to act after LCV establishment, as it does not interfere with the trafficking of the LCV [52]. Our results further indicate that the type I IFN-mediated resistance pathway affects bacterial numbers in replication vacuoles by activating bacterial killing [52]. This pathway most likely involves the IFN-stimulated GTPase IRGM1 and other genes that are type I IFN-dependently upregulated in *L. pneumophila*-infected cells [52].

During *L. pneumophila* lung infection in vivo, type I and II IFNs play a partly redundant role. Whereas mice deficient for the IFNGR have impaired bacterial clearance from the lung compared to wild-type mice, mice lacking type I IFN signaling show no defect [52]. However, mice lacking receptors for both type I and II IFNs have a strongly enhanced bacterial load after infection as compared to mice lacking only IFNGR [52]. Moreover, type I and II IFNs contribute to expression of IFN-stimulated genes in the lung during infection. Whereas some genes are dependent on either type I or II IFNs, others such as IRGM1 are regulated by both types of IFNs [52]. Although further investigations are required, it appears reasonable to assume that both types of IFNs stimulate defense against *L. pneumophila* through expression of antibacterial proteins that possibly locate to the LCV. Considering the partly redundant effects of the type I and II IFNs on *L. pneumophila* infection in vivo, these antibacterial proteins are possibly induced by both types of IFNs although IFNγ may be a stronger inducer.

Conclusion

Studies in *L. pneumophila* infection clearly show that type I IFNs can contribute to antibacterial immunity. Further research work is required to better understand how type I as well as II IFNs activate the macrophage-intrinsic antibacterial defense.

References

1. Ginevra C, Duclos A, Vanhems P, Campese C, Forey F, Lina G, Che D, Etienne J, Jarraud S (2009) Host-related risk factors and clinical features of community-acquired legionnaires disease due to the Paris and Lorraine endemic strains, 1998–2007, France. Clin Infect Dis 49: 184–191
2. von Baum H, Ewig S, Marre R, Suttorp N, Gonschior S, Welte T, Luck C; Competence Network for Community Acquired Pneumonia Study Group (2008) Community-acquired *Legionella* pneumonia: new insights from the German competence network for community acquired pneumonia. Clin Infect Dis 46:1356–1364
3. Ricketts KD, Joseph CA; European Working Group for *Legionella* Infections (2007) Legionnaires disease in Europe: 2005–2006. Euro Surveill 12:E7–E8
4. Beaute J, Robesyn E, de Jong B, European Legionnaires' Disease Surveillance Network (2013) Legionnaires' disease in Europe: all quiet on the eastern front? Eur Respir J 42: 1454–1458
5. Isberg RR, O'Connor TJ, Heidtman M (2009) The *Legionella pneumophila* replication vacuole: making a cosy niche inside host cells. Nat Rev Microbiol 7:13–24
6. Hubber A, Roy CR (2010) Modulation of host cell function by *Legionella pneumophila* type IV effectors. Annu Rev Cell Dev Biol 26:261–283
7. Archer KA, Ader F, Kobayashi KS, Flavell RA, Roy CR (2010) Cooperation between multiple microbial pattern recognition systems is important for host protection against the intracellular pathogen *Legionella pneumophila*. Infect Immun 78:2477–2487
8. Akamine M, Higa F, Arakaki N, Kawakami K, Takeda K, Akira S, Saito A (2005) Differential roles of Toll-like receptors 2 and 4 in in vitro responses of macrophages to *Legionella pneumophila*. Infect Immun 73:352–361
9. Archer KA, Roy CR (2006) MyD88-dependent responses involving toll-like receptor 2 are important for protection and clearance of *Legionella pneumophila* in a mouse model of Legionnaires' disease. Infect Immun 74:3325–3333
10. Fuse ET, Tateda K, Kikuchi Y, Matsumoto T, Gondaira F, Azuma A, Kudoh S, Standiford TJ, Yamaguchi K (2007) Role of Toll-like receptor 2 in recognition of *Legionella pneumophila* in a murine pneumonia model. J Med Microbiol 56:305–312
11. Newton CA, Perkins I, Widen RH, Friedman H, Klein TW (2007) Role of Toll-like receptor 9 in *Legionella pneumophila*-induced interleukin-12 p40 production in bone marrow-derived

dendritic cells and macrophages from permissive and nonpermissive mice. Infect Immun 75: 146–151

12. Girard R, Pedron T, Uematsu S, Balloy V, Chignard M, Akira S, Chaby R (2003) Lipopolysaccharides from *Legionella* and *Rhizobium* stimulate mouse bone marrow granulocytes via Toll-like receptor 2. J Cell Sci 116:293–302
13. Blanchard DK, Djeu JY, Klein TW, Friedman H, Stewart WE 2nd (1988) Protective effects of tumor necrosis factor in experimental *Legionella pneumophila* infections of mice via activation of PMN function. J Leukoc Biol 43:429–435
14. Fujita M, Ikegame S, Harada E, Ouchi H, Inoshima I, Watanabe K, Yoshida S, Nakanishi Y (2008) TNF receptor 1 and 2 contribute in different ways to resistance to *Legionella pneumophila-induced* mortality in mice. Cytokine 44:298–303
15. Hawn TR, Smith KD, Aderem A, Skerrett SJ (2006) Myeloid differentiation primary response gene (88)- and toll-like receptor 2-deficient mice are susceptible to infection with aerosolized *Legionella pneumophila*. J Infect Dis 193:1693–1702
16. Bhan U, Trujillo G, Lyn-Kew K, Newstead MW, Zeng X, Hogaboam CM, Krieg AM, Standiford TJ (2008) Toll-like receptor 9 regulates the lung macrophage phenotype and host immunity in murine pneumonia caused by *Legionella pneumophila*. Infect Immun 76: 2895–2904
17. Hawn TR, Berrington WR, Smith IA, Uematsu S, Akira S, Aderem A, Smith KD, Skerrett SJ (2007) Altered inflammatory responses in TLR5-deficient mice infected with *Legionella pneumophila*. J Immunol 179:6981–6987
18. Archer KA, Alexopoulou L, Flavell RA, Roy CR (2009) Multiple MyD88-dependent responses contribute to pulmonary clearance of *Legionella pneumophila*. Cell Microbiol 11:21–36
19. Frutuoso MS, Hori JI, Pereira MS, Junior DS, Sonego F, Kobayashi KS, Flavell RA, Cunha FQ, Zamboni DS (2010) The pattern recognition receptors Nod1 and Nod2 account for neutrophil recruitment to the lungs of mice infected with *Legionella pneumophila*. Microbes Infect 12:819–827
20. Shin S, Case CL, Archer KA, Nogueira CV, Kobayashi KS, Flavell RA, Roy CR, Zamboni DS (2008) Type IV secretion-dependent activation of host MAP kinases induces an increased proinflammatory cytokine response to *Legionella pneumophila*. PLoS Pathog 4:e1000220
21. Diez E, Lee SH, Gauthier S, Yaraghi Z, Tremblay M, Vidal S, Gros P (2003) Birc1e is the gene within the Lgn1 locus associated with resistance to *Legionella pneumophila*. Nat Genet 33: 55–60
22. Wright EK, Goodart SA, Growney JD, Hadinoto V, Endrizzi MG, Long EM, Sadigh K, Abney AL, Bernstein-Hanley I, Dietrich WF (2003) Naip5 affects host susceptibility to the intracellular pathogen *Legionella pneumophila*. Curr Biol 13:27–36
23. Poyet JL, Srinivasula SM, Tnani M, Razmara M, Fernandes-Alnemri T, Alnemri ES (2001) Identification of Ipaf, a human caspase-1-activating protein related to Apaf-1. J Biol Chem 276:28309–28313
24. Zamboni DS, Kobayashi KS, Kohlsdorf T, Ogura Y, Long EM, Vance RE, Kuida K, Mariathasan S, Dixit VM, Flavell RA, Dietrich WF, Roy CR (2006) The Birc1e cytosolic pattern-recognition receptor contributes to the detection and control of *Legionella pneumophila* infection. Nat Immunol 7:318–325
25. Mariathasan S, Newton K, Monack DM, Vucic D, French DM, Lee WP, Roose-Girma M, Erickson S, Dixit VM (2004) Differential activation of the inflammasome by caspase-1 adaptors ASC and Ipaf. Nature 430:213–218
26. Lightfield KL, Persson J, Trinidad NJ, Brubaker SW, Kofoed EM, Sauer JD, Dunipace EA, Warren SE, Miao EA, Vance RE (2011) Differential requirements for NAIP5 in activation of the NLRC4 inflammasome. Infect Immun 79:1606–1614
27. Pereira MS, Morgantetti GF, Massis LM, Horta CV, Hori JI, Zamboni DS (2011) Activation of NLRC4 by flagellated bacteria triggers caspase-1-dependent and -independent responses to restrict *Legionella pneumophila* replication in macrophages and in vivo. J Immunol 187: 6447–6455

28. Molofsky AB, Byrne BG, Whitfield NN, Madigan CA, Fuse ET, Tateda K, Swanson MS (2006) Cytosolic recognition of flagellin by mouse macrophages restricts *Legionella pneumophila* infection. J Exp Med 203:1093–1104
29. Ren T, Zamboni DS, Roy CR, Dietrich WF, Vance RE (2006) Flagellin-deficient *Legionella* mutants evade caspase-1- and Naip5-mediated macrophage immunity. PLoS Pathog 2:e18
30. Amer A, Franchi L, Kanneganti TD, Body-Malapel M, Ozoren N, Brady G, Meshinchi S, Jagirdar R, Gewirtz A, Akira S, Nunez G (2006) Regulation of *Legionella* phagosome maturation and infection through flagellin and host Ipaf. J Biol Chem 281:35217–35223
31. Fortier A, de Chastellier C, Balor S, Gros P (2007) Birc1e/Naip5 rapidly antagonizes modulation of phagosome maturation by *Legionella pneumophila*. Cell Microbiol 9:910–923
32. Casson CN, Copenhaver AM, Zwack EE, Nguyen HT, Strowig T, Javdan B, Bradley WP, Fung TC, Flavell RA, Brodsky IE, Shin S (2013) Caspase-11 activation in response to bacterial secretion systems that access the host cytosol. PLoS Pathog 9:e1003400
33. Case CL, Shin S, Roy CR (2009) Asc and Ipaf inflammasomes direct distinct pathways for caspase-1 activation in response to *Legionella pneumophila*. Infect Immun 77:1981–1991
34. Aachoui Y, Leaf IA, Hagar JA, Fontana MF, Campos CG, Zak DE, Tan MH, Cotter PA, Vance RE, Aderem A, Miao EA (2013) Caspase-11 protects against bacteria that escape the vacuole. Science 339:975–978
35. Case CL, Kohler LJ, Lima JB, Strowig T, de Zoete MR, Flavell RA, Zamboni DS, Roy CR (2013) Caspase-11 stimulates rapid flagellin-independent pyroptosis in response to *Legionella pneumophila*. Proc Natl Acad Sci U S A 110:1851–1856
36. Akhter A, Caution K, Abu Khweek A, Tazi M, Abdulrahman BA, Abdelaziz DH, Voss OH, Doseff AI, Hassan H, Azad AK, Schlesinger LS, Wewers MD, Gavrilin MA, Amer AO (2012) Caspase-11 promotes the fusion of phagosomes harboring pathogenic bacteria with lysosomes by modulating actin polymerization. Immunity 37:35–47
37. Fontana MF, Banga S, Barry KC, Shen X, Tan Y, Luo ZQ, Vance RE (2011) Secreted bacterial effectors that inhibit host protein synthesis are critical for induction of the innate immune response to virulent *Legionella pneumophila*. PLoS Pathog 7:e1001289
38. Ivanov SS, Roy CR (2013) Pathogen signatures activate a ubiquitination pathway that modulates the function of the metabolic checkpoint kinase mTOR. Nat Immunol 14:1219–1228
39. Barry KC, Fontana MF, Portman JL, Dugan AS, Vance RE (2013) IL-1alpha signaling initiates the inflammatory response to virulent *Legionella pneumophila* in vivo. J Immunol 190: 6329–6339
40. LeibundGut-Landmann S, Weidner K, Hilbi H, Oxenius A (2011) Nonhematopoietic cells are key players in innate control of bacterial airway infection. J Immunol 186:3130–3137
41. Horwitz MA, Silverstein SC (1981) Interaction of the Legionnaires' disease bacterium (*Legionella pneumophila*) with human phagocytes. I. *L. pneumophila* resists killing by polymorphonuclear leukocytes, antibody, and complement. J Exp Med 153:386–397
42. Weinbaum DL, Bailey J, Benner RR, Pasculle AW, Dowling JN (1983) The contribution of human neutrophils and serum to host defense against *Legionella micdadei*. J Infect Dis 148: 510–517
43. Tateda K, Moore TA, Deng JC, Newstead MW, Zeng X, Matsukawa A, Swanson MS, Yamaguchi K, Standiford TJ (2001) Early recruitment of neutrophils determines subsequent T1/T2 host responses in a murine model of *Legionella pneumophila* pneumonia. J Immunol 166:3355–3361
44. Sporri R, Joller N, Hilbi H, Oxenius A (2008) A novel role for neutrophils as critical activators of NK cells. J Immunol 181:7121–7130
45. Blanchard DK, Friedman H, Stewart WE 2nd, Klein TW, Djeu JY (1988) Role of gamma interferon in induction of natural killer activity by *Legionella pneumophila* in vitro and in an experimental murine infection model. Infect Immun 56:1187–1193
46. Heath L, Chrisp C, Huffnagle G, LeGendre M, Osawa Y, Hurley M, Engleberg C, Fantone J, Brieland J (1996) Effector mechanisms responsible for gamma interferon-mediated host resistance to *Legionella pneumophila* lung infection: the role of endogenous nitric oxide differs in susceptible and resistant murine hosts. Infect Immun 64:5151–5160

47. Shinozawa Y, Matsumoto T, Uchida K, Tsujimoto S, Iwakura Y, Yamaguchi K (2002) Role of interferon-gamma in inflammatory responses in murine respiratory infection with *Legionella pneumophila*. J Med Microbiol 51:225–230
48. Sporri R, Joller N, Albers U, Hilbi H, Oxenius A (2006) MyD88-dependent IFN-gamma production by NK cells is key for control of *Legionella pneumophila* infection. J Immunol 176:6162–6171
49. Ang DK, Oates CV, Schuelein R, Kelly M, Sansom FM, Bourges D, Boon L, Hertzog PJ, Hartland EL, van Driel IR (2010) Cutting edge: pulmonary *Legionella pneumophila* is controlled by plasmacytoid dendritic cells but not type I IFN. J Immunol 184:5429–5433
50. Stetson DB, Medzhitov R (2006) Recognition of cytosolic DNA activates an IRF3-dependent innate immune response. Immunity 24:93–103
51. Opitz B, Vinzing M, van Laak V, Schmeck B, Heine G, Gunther S, Preissner R, Slevogt H, N'Guessan PD, Eitel J, Goldmann T, Flieger A, Suttorp N, Hippenstiel S (2006) *Legionella pneumophila* induces IFNbeta in lung epithelial cells via IPS-1 and IRF3, which also control bacterial replication. J Biol Chem 281:36173–36179
52. Lippmann J, Muller HC, Naujoks J, Tabeling C, Shin S, Witzenrath M, Hellwig K, Kirschning CJ, Taylor GA, Barchet W, Bauer S, Suttorp N, Roy CR, Opitz B (2011) Dissection of a type I interferon pathway in controlling bacterial intracellular infection in mice. Cell Microbiol 13:1668–1682
53. Vogel JP, Andrews HL, Wong SK, Isberg RR (1998) Conjugative transfer by the virulence system of *Legionella pneumophila*. Science 279:873–876
54. Creasey EA, Isberg RR (2012) The protein SdhA maintains the integrity of the *Legionellaontaining* vacuole. Proc Natl Acad Sci U S A 109:3481–3486
55. Monroe KM, McWhirter SM, Vance RE (2009) Identification of host cytosolic sensors and bacterial factors regulating the type I interferon response to *Legionella pneumophila*. PLoS Pathog 5:e1000665
56. Ge J, Gong YN, Xu Y, Shao F (2012) Preventing bacterial DNA release and absent in melanoma 2 inflammasome activation by a *Legionella* effector functioning in membrane trafficking. Proc Natl Acad Sci U S A 109:6193–6198
57. Ishikawa H, Ma Z, Barber GN (2009) STING regulates intracellular DNA-mediated, type I interferon-dependent innate immunity. Nature 461:788–792
58. Sun W, Li Y, Chen L, Chen H, You F, Zhou X, Zhou Y, Zhai Z, Chen D, Jiang Z (2009) ERIS, an endoplasmic reticulum IFN stimulator, activates innate immune signaling through dimerization. Proc Natl Acad Sci U S A 106:8653–8658
59. Sun L, Wu J, Du F, Chen X, Chen ZJ (2013) Cyclic GMP-AMP synthase is a cytosolic DNA sensor that activates the type I interferon pathway. Science 339:786–791
60. Unterholzner L, Keating SE, Baran M, Horan KA, Jensen SB, Sharma S, Sirois CM, Jin T, Latz E, Xiao TS, Fitzgerald KA, Paludan SR, Bowie AG (2010) IFI16 is an innate immune sensor for intracellular DNA. Nat Immunol 11:997–1004
61. Takaoka A, Wang Z, Choi MK, Yanai H, Negishi H, Ban T, Lu Y, Miyagishi M, Kodama T, Honda K, Ohba Y, Taniguchi T (2007) DAI (DLM-1/ZBP1) is a cytosolic DNA sensor and an activator of innate immune response. Nature 448:501–505
62. Zhang Z, Yuan B, Bao M, Lu N, Kim T, Liu YJ (2011) The helicase DDX41 senses intracellular DNA mediated by the adaptor STING in dendritic cells. Nat Immunol 12:959–965
63. Ablasser A, Bauernfeind F, Hartmann G, Latz E, Fitzgerald KA, Hornung V (2009) RIG-I-dependent sensing of poly(dA:dT) through the induction of an RNA polymerase III-transcribed RNA intermediate. Nat Immunol 10:1065–1072
64. Chiu YH, Macmillan JB, Chen ZJ (2009) RNA polymerase III detects cytosolic DNA and induces type I interferons through the RIG-I pathway. Cell 138:576–591
65. Lippmann J, Rothenburg S, Deigendesch N, Eitel J, Meixenberger K, van Laak V, Slevogt H, N'Guessan PD, Hippenstiel S, Chakraborty T, Flieger A, Suttorp N, Opitz B (2008) IFNbeta responses induced by intracellular bacteria or cytosolic DNA in different human cells do not require ZBP1 (DLM-1/DAI). Cell Microbiol 10:2579–2588

66. Burdette DL, Monroe KM, Sotelo-Troha K, Iwig JS, Eckert B, Hyodo M, Hayakawa Y, Vance RE (2011) STING is a direct innate immune sensor of cyclic di-GMP. Nature 478:515–518
67. Jin L, Hill KK, Filak H, Mogan J, Knowles H, Zhang B, Perraud AL, Cambier JC, Lenz LL (2011) MPYS is required for IFN response factor 3 activation and type I IFN production in the response of cultured phagocytes to bacterial second messengers cyclic-di-AMP and cyclic-di-GMP. J Immunol 187:2595–2601
68. Abdul-Sater AA, Grajkowski A, Erdjument-Bromage H, Plumlee C, Levi A, Schreiber MT, Lee C, Shuman H, Beaucage SL, Schindler C (2012) The overlapping host responses to bacterial cyclic dinucleotides. Microbes Infect 14:188–197
69. Schiavoni G, Mauri C, Carlei D, Belardelli F, Pastoris MC, Proietti E (2004) Type I IFN protects permissive macrophages from *Legionella pneumophila* infection through an IFN-gamma-independent pathway. J Immunol 173:1266–1275
70. Plumlee CR, Lee C, Beg AA, Decker T, Shuman HA, Schindler C (2009) Interferons direct an effective innate response to *Legionella pneumophila* infection. J Biol Chem 284: 30058–30066
71. Coers J, Vance RE, Fontana MF, Dietrich WF (2007) Restriction of *Legionella pneumophila* growth in macrophages requires the concerted action of cytokine and Naip5/Ipaf signalling pathways. Cell Microbiol 9:2344–2357

Type I Interferons in Immune Defense Against Streptococci

Pavel Kovarik, Virginia Castiglia, and Marton Janos

Type I Interferons in Bacterial Infections

The function of type I interferons (IFNs) in viral infections is well established and can be almost uniformly described as protective. In contrast, their role in the context of bacterial infections is much less clear, as both beneficial and detrimental effects of type I IFN signaling have been reported in animal models [1, 2]. Examples where type I IFNs confer a protective role can be found in cases of infection with *Salmonella typhimurium*, Group B Streptococcus (GBS), *Legionella pneumophila*, and *Streptococcus pneumoniae* [3–6]. The molecular mechanisms underlying type I IFN function in the context of these infections range from the induction of cytokines and iNOS, to the enhanced differentiation of inflammatory macrophages, and may also include more complex processes, which orchestrate innate and adaptive immune responses. On the other hand, in cases of infection with *Listeria monocytogenes* and *Francisella tularensis*, type I IFNs exert unfavorable functions [7–11]. Various mechanisms can explain these harmful effects, such as type I IFN-mediated apoptosis of infected lymphocytes and macrophages, IFN-dependent reduction of IL-17 production by γδT cells, or diminished neutrophil activity. In summary, it is currently not possible to identify the denominator of either beneficial or detrimental effects of type I IFNs. Given the profound effects of these immunomodulatory cytokines on the outcome of bacterial infections, elucidating their incompletely understood induction by bacteria is of immense importance [12]. In the following, we will review the current understanding of the role of type I IFNs, as well as of the mechanisms of their induction in host defense against *Streptococcus pyogenes* (Group A Streptococcus, GAS), *Streptococcus agalactiae* (GBS), and *S. pneumoniae*.

P. Kovarik (✉) • V. Castiglia • M. Janos
Max F. Perutz Laboratories, University of Vienna, Dr. Bohr-Gasse 9, 1030 Vienna, Austria
e-mail: pavel.kovarik@univie.ac.at

© Springer International Publishing Switzerland 2014

D. Parker (ed.), *Bacterial Activation of Type I Interferons*,
DOI 10.1007/978-3-319-09498-4_4

These streptococcal species are major human pathogens which, despite a long history of intense research, continue to pose a serious health threat worldwide. In this context, new therapies employing modulation of cytokine activities are attractive yet underexplored strategies.

Group A Streptococcus (*S. pyogenes*)

Pathogenicity

GAS, also called *S. pyogenes*, is a leading Gram-positive human pathogen. GAS causes a broad range of mostly self-limiting diseases including pharyngitis (strep throat), scarlet fever or impetigo [13, 14]. It may however establish invasive and life-threatening infections, such as necrotizing fasciitis and toxic shock, which result in mortality rates of more than 30 % [13]. GAS accounts for over 700 million mild and more than 650,000 severe invasive infections worldwide annually [15]. GAS and *S. pneumoniae* are the most frequently found coinfecting bacteria in specimens of the 1918 influenza pandemic and in patients of the recent H1N1 influenza outbreak [16, 17]. Analysis of patient samples and animal studies reveal that the exceptionally wide range of GAS-caused diseases along with the transition from contained to invasive infections is determined by the virulence factor armament of a particular bacterial strain and by the genetic inventory of the host immune system [13, 18–20]. The underlying host–pathogen interactions are not well understood. Virulence factors include T cell-activating superantigens, surface-localized proteins such as the serotype-determining M protein interfering with the complement system and phagocytosis, the internalization-inhibiting hyaluronic acid capsule, secreted proteases with cytokine/chemokine-inactivating properties, secreted DNases that help bacterial dissemination, and the cytolysins SLO and SLS [19, 21–23]. Horizontal bacteriophage-mediated genetic transfer and the counteracting CRISPR system contribute to the virulence diversity observed between GAS strains [24–26]. On the host side, animal studies demonstrated that innate immune cells, most notably macrophages, dendritic cells, and neutrophils, play an essential role in successful defense during subcutaneous infection, a model of invasive GAS infection [27–29]. In models of upper respiratory tract infections, mucosal Th17 cells have been found to exert protective effects although the specific effector function of these cells in GAS infections remain to be identified [30, 31]. IL-17-mediated activation of antibacterial innate immune mechanisms could be involved in the Th17-dependent defense. Interestingly, in mice the variability of individual innate immune responses contributes to differences in susceptibility to GAS infections more than the variability in T cell-mediated responses [32].

Despite the fact that GAS is a human-specific pathogen, and mice are resistant against GAS outside of laboratory conditions [33, 34], animal infection models are invaluable for understanding GAS diseases and improvements of current therapies. Consistently, much of what is known about host–pathogen interactions in GAS infections has been established from studies using gene-targeted mice. In future studies, the use of humanized mice [35] will be helpful for functional and mechanistic assessment of GAS virulence factors that target human but not murine defense systems.

Type I IFN Induction

GAS activates type I IFN production by both human and mouse innate immune cells [4, 36–38]. In addition, GAS infection of primary human macrophages triggers an IFN signaling signature resulting, among others, in the activation of the transcription factor STAT1 [37]. This signaling signature is prevented by antibodies neutralizing IFN-α and IFN-β; however, the precise nature of type I IFNs induced by GAS in human cells remains unclear. In mice, primary bone marrow-derived macrophages (BMDMs) and conventional dendritic cells (cDCs), but not plasmacytoid dendritic cells (pDCs), were shown to produce IFN-β upon GAS infection [36, 38]. In fact, GAS-derived DNA triggers IFN-β in macrophages, whereas GAS RNA stimulates IFN-β in cDCs [38] (Table 1). Generally, IFN-β is the primary type I IFN

Table 1 Ligands, host cell signaling proteins, and cell types inducing type I IFNs in streptococcal infections

Pathogen	Ligand	Signaling proteins	Host cells	References
GAS	DNA RNA Live bacteria Live bacteria	MyD88, STING, TBK1, IRF3 MyD88, IRF5 TLR7, MyD88, IRF1 STAT1, IRF1, MxA	BMDMs (mice) cDCs (mice) cDCs (mice) Human primary macrophages	[36, 38] [36, 38] [4] [37]
GBS	DNA RNA	TBK1, IRF3 TLR7, MyD88	BMDMs (mice) Peritoneal macrophages (mice) cDCs (mice)	[75] [74] [4]
S. pneumoniae	DNA	DAI, TBK1, STING, IRF3	Nasal epithelial cells, epithelial cell of the respiratory tract cDCs (mice) Nasal lymphoid associated tissues (mice) Alveolar macrophages (humans, mice), BMDMs (mice)	[6, 55] [91] [92]

IFN interferon, *MyD88* myeloid differentiation primary-response protein 88, *TBK1* TANK-binding kinase, *STING* stimulator of IFN genes, *IRF* IFN regulatory factor, *STAT1* signal inducer and activator of transcription 1, *TLR7* toll-like receptor 7, *DAI* DNA-dependent activator of IRFs, *BMDMs* bone marrow-derived macrophages, *cDCs* conventional dendritic cells

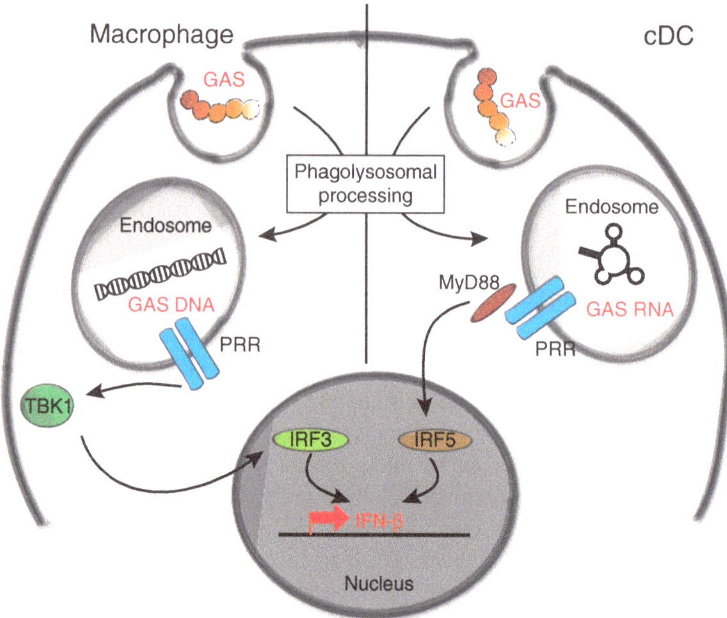

Fig. 1 Type IFN signaling and induction by GAS. GAS-derived DNA induces IFN-β in macrophages in a TBK1- and IRF3-dependent way. GAS-derived RNA induces IFN-β in cDCs via MyD88 and IRF5. Both pathways require functional phagocytosis and endosomal signaling

produced upon infection—it up-regulates the transcription factor IRF7 which then triggers IFN-α genes [39].

GAS-induced IFN-β activates the transcription factor STAT1 and STAT1 target genes in an IFNAR (type I IFN receptor)-dependent manner, confirming and clarifying a functional involvement of IFN signaling downstream of type I IFN production [36]. In contrast, the mechanism of type I IFN induction by GAS is incompletely understood (Fig. 1). Most importantly, the pattern recognition receptors (PRRs) triggering the IFN-β gene are not known [38]. In general, the identity of PRRs able to sense GAS remains one of the most challenging questions. The sole involvement of TLR2, the PRR recognizing cell wall components of Gram-positive bacteria, as well as of TLR1, TLR4, and TLR6 have been excluded [36, 38, 40]. Similarly, nucleic acid-recognizing TLR3, TLR7, and TLR9 are dispensable for production of inflammatory cytokines and type I IFNs by GAS-infected innate immune cells [36, 38, 40]. Further, type I IFNs are induced independently of the cytosolic PRRs NOD1 and NOD2 [38], which were shown to be required for IFN-β stimulation in several viral and bacterial infection models [41, 42]. Attempts at identifying the proximal GAS sensor have been performed employing cells derived from mice deficient in multiple TLRs. TLR2/TLR4 and TLR2/TLR6 double-deficient BMDMs and cDCs were not impaired in GAS recognition. It remains to be elucidated whether and how the newly characterized TLR13 is involved in GAS recognition and type I IFN induction. TLR13 is activated by a conserved sequence within the 23S rRNA of both Gram-negative and -positive bacteria [43, 44]. TLR13 stimulation causes production of inflammatory cytokines including TNF, IL-6, and IL-1β, but its role in type I IFN induction has not been clarified yet. Similarly, the role of TLR13 in host defense against bacterial pathogens remains to be investigated despite the ability of this PRR to recognize RNA of important pathogens such as *S. aureus* or GAS [43, 45]. The fact that TLR13 is expressed in mice but not humans raises the question whether humans possess an alternative route of bacterial RNA recognition. Yet another receptor that could potentially play a role in type I IFN induction by GAS is the recently characterized cyclic GMP-AMP synthase (cGAS) which acts as a cytosolic DNA sensor [46, 47]. cGAS is a danger recognition receptor which upon binding to DNA synthesizes the second messenger cyclic GMP-AMP (cGAMP). cGAMP binds and activates the ER protein STING to trigger IRF3 and IFN-β gene expression [48, 49]. While cGAS is involved in cellular defense against viruses [50–53], a role of cGAS in bacterial infections and/or in induction of type I IFNs by bacteria has not been demonstrated yet.

Signaling events downstream of the type I IFN-inducing GAS-specific PRRs are better understood (Fig. 1 and Table 1). Activation of *Ifnb* gene expression by GAS-derived DNA in macrophages is dependent on the TBK1 kinase and the transcription factor IRF3 [38]. In contrast, the IFN-β-inducing pathway triggered by GAS RNA in cDCs requires the adaptor MyD88 as well as the transcription factor IRF5, but not IRF3 [38]. Uptake of GAS is needed for triggering IFN-β production suggesting that phagolysosomal processing of internalized GAS liberates the bacterial IFN-β inducers. Whether both BMDMs and cDCs are involved in IFN-β production in vivo and whether these cell types play a redundant or distinct roles have yet to be examined.

Type I IFN Functions

Mice lacking the type I IFN receptor IFNAR1 are more susceptible to subcutaneous GAS infection [38], a standard model of severe invasive cellulitis [20]. The mortality rate of GAS-infected IFNAR1-deficient mice is 70 % whereas it is only 25 % in WT mice. IFNAR1 knockouts were shown to exhibit increased recruitment of neutrophils to the site of infection but the molecular and cellular basis of the beneficial effects of type I IFNs in GAS infection remain to be elucidated. The high neutrophil number observed in mice lacking type I IFN signaling is consistent with previous observations demonstrating inhibitory effects of type I IFNs on macrophage production of the chemokines CXCL1, CXCL2, and CCL2 during *S. pneumoniae* infections [54, 55]. These chemokines play a key role in attracting neutrophils to the site of infection. It is at present unclear how the increased neutrophil recruitment in GAS-infected IFNAR knockout mice could evoke more detrimental disease.

One can speculate that an exaggerated inflammatory response elicited by recruited neutrophils causes severe tissue damage, thereby allowing better dissemination of the pathogen. Such scenario is conceivable as GAS expresses several DNases that help liberate it from neutrophil extracellular traps (NETs) [56, 57]. Consistently, the DNase *Sda*1 is a potent virulence factor which promotes GAS to acquire an invasive infection phenotype [58]. GAS exhibits a profound propensity to induce NETs, structures that contain large amounts of inflammation-promoting material such as neutrophil DNA, histones, and other chromatin-associated proteins [59, 60]. Interestingly, TLR9, a PRR able to sense self DNA [61], might be involved in sensing GAS-induced NETs as it is beneficial in an intraperitoneal model of GAS infection [62]. This indirect role of TLR9 in GAS infections is supported by the lack of effect of TLR9 knockout on direct GAS recognition by BMDMs and cDCs [4, 38, 40]. Thus, the enhanced neutrophil recruitment in IFNAR1-deficient mice might result in more intense, hence lethal inflammation. Effects of type I IFNs on other immune reactions such as recruitment of macrophages by GAS-induced TNF [63], or IL-1β production by the GAS-activated NLRP3 inflammasome [64], should be addressed in future studies to reveal the precise role of type I IFN signaling.

Group B Streptococcus (*S. agalactiae*)

Pathogenicity

GBS, also called *S. agalactiae*, is a Gram-positive human pathogen and leading infectious agent in neonatal sepsis worldwide [65]. Neonatal sepsis causes over two million deaths annually, with decreasing incidence largely due to improved prophylactic measures [66]. In early onset neonatal disease (within 6 days after birth), GBS is transmitted vertically from mothers vaginally colonized by the pathogen. In late onset disease (7–89 days after birth), GBS infection is usually a consequence of horizontal transfer in communities. GBS is also a significant cause of maternal morbidity (bacteremia, endometritis) [67]. GBS virulence factors include the polysaccharide capsule, membrane damaging exotoxins, and adherence molecules which enable evasion of the immune system and colonization of the host [68]. Innate immune system-derived TNF, IL-1β, and nitric oxide are key defense factors in host protection [69–71]. The vulnerability of neonates to GBS results in part from underdeveloped adaptive immunity but more importantly from deficiencies in innate immunity, including limited capacity of neutrophil production and increased risk of bone marrow exhaustion [67, 72]. The neonate immune insufficiency allows colonization and infection by GBS resulting mostly in meningitis or pneumonia. Prophylactic vaccination and immunomodulation appear the most promising approaches to eradicate GBS disease [67, 73].

Type I IFN Induction and Function

Type I IFN signaling has a protective function in GBS infections: mice deficient in either type I IFN receptor or IFN-β exhibit increased mortality in a neonatal infection model, both after intravenous or intraperitoneal GBS administration [74]. This lethal infection outcome is caused by uncontrolled bacteremia, suggesting that type I IFN signaling is required for launching a complete immune and antibacterial response. Both macrophages and cDCs, but not pDCs, were identified as the source of type I IFNs [4, 74, 75] (Table 1). A direct comparison of type I IFN amounts indicate that cDCs are the major producers in vitro [4] but the principle type I IFN-producing cell in vivo has yet to be confirmed. Type I IFN production is dependent on uptake and phagolysosomal processing of GBS [4, 75] (Fig. 2). In macrophages, GBS DNA was identified as type I IFN inducer that acts along the TBK1 and IRF3 axis [75] (Fig. 2 and Table 1). GBS DNA was proposed to escape phagosomes into the cytosol where it is detected by an unknown cytosolic DNA receptor, which is different from the double-stranded DNA sensor DAI [75, 76]. The inducer of type I IFNs in cDCs was shown to be GBS RNA, which was sensed in a MyD88-dependent manner in phagosomes of infected cells [4] (Fig. 2 and Table 1). The endosomal

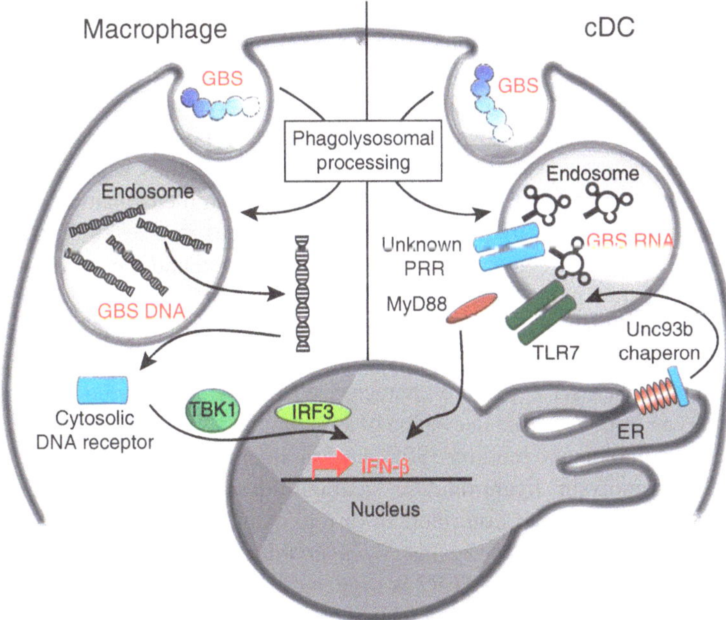

Fig. 2 Type IFN signaling and induction by GBS. Induction of type I IFNs by GBS requires uptake and phagolysosomal processing of the pathogen. In macrophages, GBS-derived DNA triggers a cytosolic sensor which signals via TBK1 and IRF3 to induce IFN-β gene expression. In cDCs, GAS-derived RNA triggers the in Unc93b-dependent way the endosomal TLR7 which signals via MyD88 toward the IFN-β gene

TLR7 was found to be involved in sensing of GBS RNA. Interestingly, GBS RNA was reported to induce TNF in macrophages independently of TLR3, TLR7, and TLR8, but it required MyD88 [77]. This RNA recognition occurs in endosomal compartments as it is dependent on Unc93b, a chaperon fundamentally involved in trafficking of endosomal TLRs. Together, these studies indicate that recognition of GBS is cell type-specific, and that GBS RNA induces type I IFNs in cDCs but not in macrophages. The molecular basis of the different outcome of GBS RNA sensing in macrophages and cDCs remains to be deciphered. As is the case with GAS, the analysis of the recently identified sensors TLR13 and cGAS might be helpful in resolving the open questions.

Streptococcus pneumoniae

Pathogenesis

S. pneumoniae (pneumococcus) is a Gram-positive human pathogen regarded as the most frequent cause of community-acquired pneumonia [78, 79]. Pneumonia is the leading lethal infectious disease in developed countries [78, 79]. *S. pneumoniae* is one of the most prominent examples of a human-specific commensal microbe that frequently turns into an infectious agent. *S. pneumoniae* asymptomatically colonizes the nasopharynx in up to 60 % of all preschool children. Yet, *S. pneumoniae* represents the prime bacterial killer among children below the age of 5 with 1.2 million deaths annually worldwide. *S. pneumoniae* also poses a serious health risk to elderly people as a consequence of age-related immunosenescence. Of particular importance is a secondary *S. pneumoniae* infection of influenza patients, *S. pneumoniae* is one of the most frequent coinfecting pathogens in cases of influenza outbreaks [16, 17]. Both the genetic makeup of the pathogen and the condition of the host immune system play decisive roles in the transition from a commensal microbe into invasive pathogen. However, the exact parameters regulating this shift are not well understood. *S. pneumoniae* occurs in more than 90 serotypes which differ in their virulence. The serotypes are characterized by their polysaccharide capsule, which plays an important role in evasion of the immune system by inhibiting phagocytosis and complement binding [80]. An armament of other virulence factors including pneumolysin, hyaluronidase, neuraminidase, the serine protease PrtA, cholin-binding proteins, etc. contribute to various extents to pneumococcal diseases [81, 82]. The immune response against *S. pneumoniae* is initiated by its interactions with innate immune receptors. TLR2 is triggered by *S. pneumoniae* cell wall components (e.g., LTA), TLR4 can be activated by pneumolysin and TLR9 recognizes pneumococcal DNA [80, 83–86]. Furthermore, the cytosolic receptors NLRP3, NOD2, and AIM2 contribute to *S. pneumoniae*-induced inflammatory cytokine induction [80, 87–90].

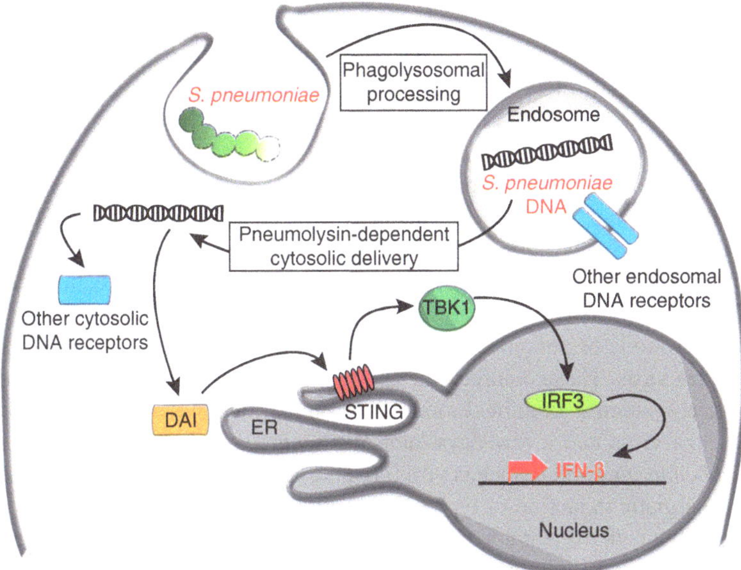

Fig. 3 Type IFN signaling and induction by *S. pneumoniae*. The cytosolic DNA sensor DAI and other cytosolic DNA receptors are involved in the induction of the IFN-β by *S. pneumoniae*. Induction of IFN-β is dependent on STING, TBK1, and IRF3

Type I IFN Induction

S. pneumoniae induces type I IFNs in nasal-associated lymphoid and epithelial tissues, as well as in human and mouse alveolar macrophages and mouse BMDMs [6, 55, 91, 92]. The type I IFN inducer is *S. pneumoniae* DNA, which is recognized upon internalization of the pathogen and/or pneumolysin-dependent cytosolic delivery [6] (Table 1). The double-stranded DNA sensor DAI participates in the detection of *S. pneumoniae* DNA as DAI-deficient cells produce less IFN-β than control cells [6] (Fig. 3). Similar to GAS and GBS, the signaling pathway downstream of the proximal sensor includes TBK1, STING and IRF3, and is possibly indirectly dependent on NOD2 [6, 55] (Fig. 3 and Table 1). Signal transduction toward the IFN-β gene proceeds in the absence of TLR4, MyD88, NOD2, and TRIF. Thus, the IFN-β-inducing properties of *S. pneumoniae*-derived DNA resemble those of GAS and GBS. It remains to be investigated whether *S. pneumoniae* RNA also possesses immunostimulatory capabilities as described for GAS and GBS.

Type I IFN Function

Intravenous infection of type I IFN signaling-deficient mice with *S. pneumoniae* results in increased lethality [74]. Further evidence for a beneficial role of type I IFNs was provided by a study using a more natural route of infection, i.e., intranasal [6]. This particular study reported an impaired clearance of the pathogen from the site of infection, i.e., from the upper respiratory tract, in mice lacking IFNAR1, despite more potent recruitment of monocytes and dendritic cells. The exact mechanism of how type I IFNs elicit protective effects in pneumococcal infections remains to be characterized.

A distinct mode of pneumococcal infection is represented by coinfections with the influenza virus. These coinfections exhibit high morbidity and are life threatening in elderly patients. In animal models of coinfections, mice are first exposed to the influenza virus and a few days later *S. pneumoniae* is delivered intranasally. Both, *S. pneumoniae* and influenza virus are able to induce type I IFNs. Coinfections lead to synergistic induction of type I IFNs and, remarkably, this high level of type I IFN signaling is detrimental to the host [54, 55, 93]. The mechanisms of the harmful effects of type I IFNs on post-influenza bacterial infection include decreased production of the chemokines CCL2, CXCL1, and CXCL2, which act as chemoattractants for monocytes and neutrophils. As a result, less monocytes and neutrophils are recruited to infected tissues, although the precise nature of the most affected leukocytes is a matter of debate [54, 55]. Further studies are needed to clarify the molecular principles of coinfections. Such studies should particularly address the inability to tolerate tissue damage, which has recently been reported to play a critical role in influenza and *L. pneumophila* coinfections [94].

Type I Interferons in Streptococcal Infections: Unifying Themes and Divergences

Although they share several common features, GAS, GBS, and *S. pneumoniae* cause diverse diseases in humans. They are Gram-positive encapsulated pathogens exhibiting a largely extracellular life cycle. Their key virulence factors are cytolysins, which possess cytotoxic properties and promote intracellular survival and/or phagolysosomal damage. These pathogens' ability to survive and grow within infected cells is very limited, although it has been reported that GAS is capable of acquiring a significant intracellular life span [13, 95]. Nonetheless, most internalized GAS are efficiently killed by the host phagolysosomal lytic and oxidative mechanisms. GAS that has escaped from the hostile phagosomal environment is rapidly recognized in the cytosol by the autophagy machinery and eradicated [96, 97]. The highly successful destruction of streptococci in the phagosomes results in the release of, among others, bacterial nucleic acids, which can act as type I IFN inducers. Consequently, endosomal recognition of GAS and GBS RNA induces type I IFNs [4, 38]. In this context, the role of *S. pneumoniae* RNA has yet to be investigated. In contrast, all three streptococcal species have been reported to induce type I IFNs by their DNA, which is sensed by cytosolic DNA receptors [6, 38, 75]. Cytolysins are likely to be involved in the passage of DNA through the phagosomal membrane, but the precise mechanisms of streptococcal DNA delivery into the host cell cytosol remain unclear. The issue of type I IFN-inducing receptors also requires further investigation. Whereas TLR7 was identified as the RNA-sensing type I IFN inducer in response to GBS but not GAS [4, 38], the DNA sensor DAI was found to induce type I IFNs in response to *S. pneumoniae* but not GBS [6, 75]. Future studies, now also include newly identified receptors, will show whether there are common type I IFN-inducing pathways in streptococcal infections.

Type I IFNs exhibit protective functions in infections against all three streptococcal species discussed here, yet the precise nature of these beneficial functions are not well explained. As the three streptococcal species cause different diseases and display in part different tissue tropism, the mode of action of type I IFNs will most likely involve multiple possibly non-overlapping mechanisms. Elucidation of type I IFN functions is essential for our better understanding of the surprisingly detrimental effects of these cytokines during viral coinfections [54, 55, 93]. Further, it has yet to be investigated whether the negative impact of type I IFNs during coinfections is restricted to respiratory pathogens.

Outlook

Despite significant advances in our understating of type I IFNs in bacterial infections, the key questions remain unresolved for most bacterial pathogens. These questions include the identity of type I IFN-inducing sensors and the specific effector functions of type I IFNs. Analyses of a broader range of innate immune receptors, ideally by employing unbiased approaches such as mass spectroscopy or genetic screens, will give us a more comprehensive picture of type I IFN induction. To elucidate the effector functions of type I IFNs, better infection models are needed. These will have to include animals allowing cell type-specific deletion of IFNAR1 [98], analysis of animals lacking different type I IFNs (particularly IFN-β), and in vivo and intravital imaging techniques. A so far unexplored aspect in streptococcal infections is the timing of type I IFN signaling. In the view of recent findings describing an unexpected harmful function of type I IFNs during persistent viral infections [99, 100], time-resolved analysis of type I IFN signaling in streptococcal infections and viral coinfections will need to be conducted in future studies. Another major challenge is the evaluation of the relevance of animal studies for the understanding of streptococcal diseases in humans. Clearly, the use of gene-targeted mice will remain fundamental for mechanistic and proof-of-principle studies. However, the increasingly better understood differences between the human and mouse immune systems, including their partially different repertoires of innate immune receptors, should be carefully considered when using animal models for human-specific pathogens.

Modulation of immune responses is recognized as a highly promising approach in the treatment of severe infectious diseases, and it may be the sole strategy for the treatment of acute life-threatening conditions such as streptococcal toxic shock syndrome. Type I IFNs are major immune modulators, possessing both immunostimulatory and immunosuppressive properties [101, 102]; as such, the elucidation of their mechanism of action in streptococcal infections could eventually establish type I IFN signaling as a target for novel therapies.

References

1. Trinchieri G (2010) Type I interferon: friend or foe? J Exp Med 207:2053–2063
2. Decker T, Muller M, Stockinger S (2005) The yin and yang of type I interferon activity in bacterial infection. Nat Rev Immunol 5:675–687
3. Freudenberg MA, Merlin T, Kalis C, Chvatchko Y, Stubig H et al (2002) Cutting edge: a murine, IL-12-independent pathway of IFN-gamma induction by gram-negative bacteria based on STAT4 activation by Type I IFN and IL-18 signaling. J Immunol 169:1665–1668
4. Mancuso G, Gambuzza M, Midiri A, Biondo C, Papasergi S et al (2009) Bacterial recognition by TLR7 in the lysosomes of conventional dendritic cells. Nat Immunol 10:587–594
5. Plumlee CR, Lee C, Beg AA, Decker T, Shuman HA et al (2009) Interferons direct an effective innate response to *Legionella pneumophila* infection. J Biol Chem 284:30058–30066
6. Parker D, Martin FJ, Soong G, Harfenist BS, Aguilar JL et al (2011) *Streptococcus pneumoniae* DNA initiates type I interferon signaling in the respiratory tract. MBio 2:e00016-00011
7. O'Connell RM, Saha SK, Vaidya SA, Bruhn KW, Miranda GA et al (2004) Type I interferon production enhances susceptibility to *Listeria monocytogenes* infection. J Exp Med 200:437–445
8. Auerbuch V, Brockstedt DG, Meyer-Morse N, O'Riordan M, Portnoy DA (2004) Mice lacking the type I interferon receptor are resistant to *Listeria monocytogenes*. J Exp Med 200: 527–533
9. Carrero JA, Calderon B, Unanue ER (2004) Type I interferon sensitizes lymphocytes to apoptosis and reduces resistance to listeria infection. J Exp Med 200:535–540
10. Stockinger S, Kastner R, Kernbauer E, Pilz A, Westermayer S et al (2009) Characterization of the interferon-producing cell in mice infected with *Listeria monocytogenes*. PLoS Pathog 5:e1000355
11. Henry T, Kirimanjeswara GS, Ruby T, Jones JW, Peng K et al (2010) Type I IFN signaling constrains IL-17A/F secretion by gammadelta T cells during bacterial infections. J Immunol 184:3755–3767
12. Monroe KM, McWhirter SM, Vance RE (2010) Induction of type I interferons by bacteria. Cell Microbiol 12:881–890
13. Johansson L, Thulin P, Low DE, Norrby-Teglund A (2010) Getting under the skin: the immunopathogenesis of *Streptococcus pyogenes* deep tissue infections. Clin Infect Dis 51:58–65
14. Wessels MR (2011) Clinical practice. Streptococcal pharyngitis. N Engl J Med 364: 648–655
15. Carapetis JR, Steer AC, Mulholland EK, Weber M (2005) The global burden of group A streptococcal diseases. Lancet Infect Dis 5:685–694
16. Morens DM, Fauci AS (2007) The 1918 influenza pandemic: insights for the 21st century. J Infect Dis 195:1018–1028

17. Zakikhany K, Degail MA, Lamagni T, Waight P, Guy R et al (2011) Increase in invasive *Streptococcus pyogenes* and *Streptococcus pneumoniae* infections in England, December 2010 to January 2011. Euro Surveill 16
18. Kotb M, Norrby-Teglund A, McGeer A, El-Sherbini H, Dorak MT et al (2002) An immunogenetic and molecular basis for differences in outcomes of invasive group A streptococcal infections. Nat Med 8:1398–1404
19. Olsen RJ, Shelburne SA, Musser JM (2009) Molecular mechanisms underlying group A streptococcal pathogenesis. Cell Microbiol 11:1–12
20. Medina E (2010) Murine model of cutaneous infection with *Streptococcus pyogenes*. Methods Mol Biol 602:395–403
21. Bisno AL, Brito MO, Collins CM (2003) Molecular basis of group A streptococcal virulence. Lancet Infect Dis 3:191–200
22. Lynskey NN, Lawrenson RA, Sriskandan S (2011) New understandings in *Streptococcus pyogenes*. Curr Opin Infect Dis 24:196–202
23. Molloy EM, Cotter PD, Hill C, Mitchell DA, Ross RP (2011) Streptolysin S-like virulence factors: the continuing sagA. Nat Rev Microbiol 9:670–681
24. Sitkiewicz I, Nagiec MJ, Sumby P, Butler SD, Cywes-Bentley C et al (2006) Emergence of a bacterial clone with enhanced virulence by acquisition of a phage encoding a secreted phospholipase A2. Proc Natl Acad Sci U S A 103:16009–16014
25. Deltcheva E, Chylinski K, Sharma CM, Gonzales K, Chao Y et al (2011) CRISPR RNA maturation by trans-encoded small RNA and host factor RNase III. Nature 471:602–607
26. Nozawa T, Furukawa N, Aikawa C, Watanabe T, Haobam B et al (2011) CRISPR inhibition of prophage acquisition in *Streptococcus pyogenes*. PLoS One 6:e19543
27. Goldmann O, Rohde M, Chhatwal GS, Medina E (2004) Role of macrophages in host resistance to group A streptococci. Infect Immun 72:2956–2963
28. Loof TG, Rohde M, Chhatwal GS, Jung S, Medina E (2007) The contribution of dendritic cells to host defenses against *Streptococcus pyogenes*. J Infect Dis 196:1794–1803
29. Navarini AA, Lang KS, Verschoor A, Recher M, Zinkernagel AS et al (2009) Innate immune-induced depletion of bone marrow neutrophils aggravates systemic bacterial infections. Proc Natl Acad Sci U S A 106:7107–7112
30. Wang B, Dileepan T, Briscoe S, Hyland KA, Kang J et al (2010) Induction of TGF-beta1 and TGF-beta1-dependent predominant Th17 differentiation by group A streptococcal infection. Proc Natl Acad Sci U S A 107:5937–5942
31. Dileepan T, Linehan JL, Moon JJ, Pepper M, Jenkins MK et al (2011) Robust antigen specific Th17 T cell response to group A *Streptococcus* is dependent on IL-6 and intranasal route of infection. PLoS Pathog 7:e1002252
32. Goldmann O, Lengeling A, Bose J, Bloecker H, Geffers R et al (2005) The role of the MHC on resistance to group a streptococci in mice. J Immunol 175:3862–3872
33. Fox GF, Anderson LC, Loew FM, Quimby FW (2002) Laboratory animal medicine. Academic, Elsevier
34. Tart AH, Walker MJ, Musser JM (2007) New understanding of the group A *Streptococcus* pathogenesis cycle. Trends Microbiol 15:318–325
35. Brehm MA, Jouvet N, Greiner DL, Shultz LD (2013) Humanized mice for the study of infectious diseases. Curr Opin Immunol 25:428–435
36. Gratz N, Siller M, Schaljo B, Pirzada ZA, Gattermeier I et al (2008) Group A *Streptococcus* activates type I interferon production and MyD88-dependent signaling without involvement of TLR2, TLR4, and TLR9. J Biol Chem 283:19879–19887
37. Miettinen M, Lehtonen A, Julkunen I, Matikainen S (2000) *Lactobacilli* and *Streptococci* activate NF-kappa B and STAT signaling pathways in human macrophages. J Immunol 164:3733–3740
38. Gratz N, Hartweger H, Matt U, Kratochvill F, Janos M et al (2011) Type I interferon production induced by *Streptococcus pyogenes*-derived nucleic acids is required for host protection. PLoS Pathog 7:e1001345

39. Marie I, Durbin JE, Levy DE (1998) Differential viral induction of distinct interferon-alpha genes by positive feedback through interferon regulatory factor-7. EMBO J 17:6660–6669
40. Loof TG, Goldmann O, Medina E (2008) Immune recognition of *Streptococcus pyogenes* by dendritic cells. Infect Immun 76:2785–2792
41. Watanabe T, Asano N, Murray PJ, Ozato K, Tailor P et al (2008) Muramyl dipeptide activation of nucleotide-binding oligomerization domain 2 protects mice from experimental colitis. J Clin Invest 118:545–559
42. Sabbah A, Chang TH, Harnack R, Frohlich V, Tominaga K et al (2009) Activation of innate immune antiviral responses by Nod2. Nat Immunol 10:1073–1080
43. Oldenburg M, Kruger A, Ferstl R, Kaufmann A, Nees G et al (2012) TLR13 recognizes bacterial 23S rRNA devoid of erythromycin resistance-forming modification. Science 337: 1111–1115
44. Li XD, Chen ZJ (2012) Sequence specific detection of bacterial 23S ribosomal RNA by TLR13. Elife 1:e00102
45. Hidmark A, von Saint Paul A, Dalpke AH (2012) Cutting edge: TLR13 is a receptor for bacterial RNA. J Immunol 189:2717–2721
46. Sun L, Wu J, Du F, Chen X, Chen ZJ (2012) Cyclic GMP-AMP synthase is a cytosolic DNA sensor that activates the type I interferon pathway. Science 339(6121):786–791
47. Wu J, Sun L, Chen X, Du F, Shi H et al (2012) Cyclic GMP-AMP is an endogenous second messenger in innate immune signaling by cytosolic DNA. Science 339(6121):826–830
48. Ishikawa H, Barber GN (2008) STING is an endoplasmic reticulum adaptor that facilitates innate immune signalling. Nature 455:674–678
49. Ishikawa H, Ma Z, Barber GN (2009) STING regulates intracellular DNA-mediated, type I interferon-dependent innate immunity. Nature 461:788–792
50. Li XD, Wu J, Gao D, Wang H, Sun L et al (2013) Pivotal roles of cGAS-cGAMP signaling in antiviral defense and immune adjuvant effects. Science 341:1390–1394
51. Gao D, Wu J, Wu YT, Du F, Aroh C et al (2013) Cyclic GMP-AMP synthase is an innate immune sensor of HIV and other retroviruses. Science 341:903–906
52. Schoggins JW, MacDuff DA, Imanaka N, Gainey MD, Shrestha B et al (2014) Pan-viral specificity of IFN-induced genes reveals new roles for cGAS in innate immunity. Nature 505:691–695
53. Ablasser A, Schmid-Burgk JL, Hemmerling I, Horvath GL, Schmidt T et al (2013) Cell intrinsic immunity spreads to bystander cells via the intercellular transfer of cGAMP. Nature 503:530–534
54. Shahangian A, Chow EK, Tian X, Kang JR, Ghaffari A et al (2009) Type I IFNs mediate development of postinfluenza bacterial pneumonia in mice. J Clin Invest 119:1910–1920
55. Nakamura S, Davis KM, Weiser JN (2011) Synergistic stimulation of type I interferons during influenza virus coinfection promotes *Streptococcus pneumoniae* colonization in mice. J Clin Invest 121:3657–3665
56. Buchanan JT, Simpson AJ, Aziz RK, Liu GY, Kristian SA et al (2006) DNase expression allows the pathogen group A *Streptococcus* to escape killing in neutrophil extracellular traps. Curr Biol 16:396–400
57. Chang A, Khemlani A, Kang H, Proft T (2011) Functional analysis of *Streptococcus pyogenes* nuclease A (SpnA), a novel group A streptococcal virulence factor. Mol Microbiol 79:1629–1642
58. Walker MJ, Hollands A, Sanderson-Smith ML, Cole JN, Kirk JK et al (2007) DNase *Sda*1 provides selection pressure for a switch to invasive group A streptococcal infection. Nat Med 13:981–985
59. Kolaczkowska E, Kubes P (2013) Neutrophil recruitment and function in health and inflammation. Nat Rev Immunol 13:159–175
60. Brinkmann V, Zychlinsky A (2007) Beneficial suicide: why neutrophils die to make NETs. Nat Rev Microbiol 5:577–582

61. Barrat FJ, Meeker T, Gregorio J, Chan JH, Uematsu S et al (2005) Nucleic acids of mammalian origin can act as endogenous ligands for Toll-like receptors and may promote systemic lupus erythematosus. J Exp Med 202:1131–1139
62. Zinkernagel AS, Hruz P, Uchiyama S, von Kockritz-Blickwede M, Schuepbach RA et al (2012) Importance of Toll-like receptor 9 in host defense against M1T1 group A *Streptococcus* infections. J Innate Immun 4:213–218
63. Mishalian I, Ordan M, Peled A, Maly A, Eichenbaum MB et al (2011) Recruited macrophages control dissemination of group A *Streptococcus* from infected soft tissues. J Immunol 187:6022–6031
64. Harder J, Franchi L, Munoz-Planillo R, Park JH, Reimer T et al (2009) Activation of the Nlrp3 inflammasome by *Streptococcus pyogenes* requires streptolysin O and NF-{kappa}B activation but proceeds independently of TLR signaling and P2X7 receptor. J Immunol 183:5823–5829
65. Dagnew AF, Cunnington MC, Dube Q, Edwards MS, French N et al (2012) Variation in reported neonatal group B streptococcal disease incidence in developing countries. Clin Infect Dis 55:91–102
66. Lozano R, Wang H, Foreman KJ, Rajaratnam JK, Naghavi M et al (2011) Progress towards millennium development goals 4 and 5 on maternal and child mortality: an updated systematic analysis. Lancet 378:1139–1165
67. Koenig JM, Keenan WJ (2009) Group B *Streptococcus* and early-onset sepsis in the era of maternal prophylaxis. Pediatr Clin North Am 56:689–708, Table of Contents
68. Henneke P, Berner R (2006) Interaction of neonatal phagocytes with group B *Streptococcus*: recognition and response. Infect Immun 74:3085–3095
69. Mancuso G, Midiri A, Beninati C, Biondo C, Galbo R et al (2004) Dual role of TLR2 and myeloid differentiation factor 88 in a mouse model of invasive group B streptococcal disease. J Immunol 172:6324–6329
70. Costa A, Gupta R, Signorino G, Malara A, Cardile F et al (2012) Activation of the NLRP3 inflammasome by group B streptococci. J Immunol 188:1953–1960
71. Deshmukh SD, Muller S, Hese K, Rauch KS, Wennekamp J et al (2012) NO is a macrophage autonomous modifier of the cytokine response to streptococcal single-stranded RNA. J Immunol 188:774–780
72. Teti G, Mancuso G, Tomasello F, Chiofalo MS (1992) Production of tumor necrosis factor-alpha and interleukin-6 in mice infected with group B streptococci. Circ Shock 38:138–144
73. Pannaraj PS, Edwards MS, Ewing KT, Lewis AL, Rench MA et al (2009) Group B streptococcal conjugate vaccines elicit functional antibodies independent of strain O-acetylation. Vaccine 27:4452–4456
74. Mancuso G, Midiri A, Biondo C, Beninati C, Zummo S et al (2007) Type I IFN signaling is crucial for host resistance against different species of pathogenic bacteria. J Immunol 178: 3126–3133
75. Charrel-Dennis M, Latz E, Halmen KA, Trieu-Cuot P, Fitzgerald KA et al (2008) TLR-independent type I interferon induction in response to an extracellular bacterial pathogen via intracellular recognition of its DNA. Cell Host Microbe 4:543–554
76. Takaoka A, Wang Z, Choi MK, Yanai H, Negishi H et al (2007) DAI (DLM-1/ZBP1) is a cytosolic DNA sensor and an activator of innate immune response. Nature 448:501–505
77. Deshmukh SD, Kremer B, Freudenberg M, Bauer S, Golenbock DT et al (2011) Macrophages recognize streptococci through bacterial single-stranded RNA. EMBO Rep 12:71–76
78. Sinclair A, Xie X, Teltscher M, Dendukuri N (2013) Systematic review and meta-analysis of a urine-based pneumococcal antigen test for diagnosis of community-acquired pneumonia caused by *Streptococcus pneumoniae*. J Clin Microbiol 51:2303–2310
79. van der Poll T, Opal SM (2009) Pathogenesis, treatment, and prevention of pneumococcal pneumonia. Lancet 374:1543–1556

80. Koppe U, Suttorp N, Opitz B (2012) Recognition of *Streptococcus pneumoniae* by the innate immune system. Cell Microbiol 14:460–466
81. Patenge N, Fiedler T, Kreikemeyer B (2013) Common regulators of virulence in streptococci. Curr Top Microbiol Immunol 368:111–153
82. Mitchell AM, Mitchell TJ (2010) *Streptococcus pneumoniae*: virulence factors and variation. Clin Microbiol Infect 16:411–418
83. Knapp S, Wieland CW, van't Veer C, Takeuchi O, Akira S et al (2004) Toll-like receptor 2 plays a role in the early inflammatory response to murine pneumococcal pneumonia but does not contribute to antibacterial defense. J Immunol 172:3132–3138
84. Schroder NW, Morath S, Alexander C, Hamann L, Hartung T et al (2003) Lipoteichoic acid (LTA) of *Streptococcus pneumoniae* and *Staphylococcus aureus* activates immune cells via Toll-like receptor (TLR)-2, lipopolysaccharide-binding protein (LBP), and CD14, whereas TLR-4 and MD-2 are not involved. J Biol Chem 278:15587–15594
85. Albiger B, Dahlberg S, Sandgren A, Wartha F, Beiter K et al (2007) Toll-like receptor 9 acts at an early stage in host defence against pneumococcal infection. Cell Microbiol 9:633–644
86. Malley R, Henneke P, Morse SC, Cieslewicz MJ, Lipsitch M et al (2003) Recognition of pneumolysin by Toll-like receptor 4 confers resistance to pneumococcal infection. Proc Natl Acad Sci U S A 100:1966–1971
87. Davis KM, Nakamura S, Weiser JN (2011) Nod2 sensing of lysozyme-digested peptidoglycan promotes macrophage recruitment and clearance of *S. pneumoniae* colonization in mice. J Clin Invest 121:3666–3676
88. Witzenrath M, Pache F, Lorenz D, Koppe U, Gutbier B et al (2011) The NLRP3 inflammasome is differentially activated by pneumolysin variants and contributes to host defense in pneumococcal pneumonia. J Immunol 187:434–440
89. McNeela EA, Burke A, Neill DR, Baxter C, Fernandes VE et al (2010) Pneumolysin activates the NLRP3 inflammasome and promotes proinflammatory cytokines independently of TLR4. PLoS Pathog 6:e1001191
90. Fang R, Tsuchiya K, Kawamura I, Shen Y, Hara H et al (2011) Critical roles of ASC inflammasomes in caspase-1 activation and host innate resistance to *Streptococcus pneumoniae* infection. J Immunol 187:4890–4899
91. Joyce EA, Popper SJ, Falkow S (2009) *Streptococcus pneumoniae* nasopharyngeal colonization induces type I interferons and interferon-induced gene expression. BMC Genomics 10:404
92. Koppe U, Hogner K, Doehn JM, Muller HC, Witzenrath M et al (2012) *Streptococcus pneumoniae* stimulates a STING- and IFN regulatory factor 3-dependent type I IFN production in macrophages, which regulates RANTES production in macrophages, cocultured alveolar epithelial cells, and mouse lungs. J Immunol 188:811–817
93. Li W, Moltedo B, Moran TM (2012) Type I interferon induction during influenza virus infection increases susceptibility to secondary *Streptococcus pneumoniae* infection by negative regulation of gammadelta T cells. J Virol 86:12304–12312
94. Jamieson AM, Pasman L, Yu S, Gamradt P, Homer RJ et al (2013) Role of tissue protection in lethal respiratory viral-bacterial coinfection. Science 340(6137):1230–1234
95. Hertzen E, Johansson L, Kansal R, Hecht A, Dahesh S et al (2012) Intracellular *Streptococcus pyogenes* in human macrophages display an altered gene expression profile. PLoS One 7:e35218
96. Nakagawa I, Amano A, Mizushima N, Yamamoto A, Yamaguchi H et al (2004) Autophagy defends cells against invading group A *Streptococcus*. Science 306:1037–1040
97. Yamaguchi H, Nakagawa I, Yamamoto A, Amano A, Noda T et al (2009) An initial step of GAS-containing autophagosome-like vacuoles formation requires Rab7. PLoS Pathog 5:e1000670
98. Prinz M, Schmidt H, Mildner A, Knobeloch KP, Hanisch UK et al (2008) Distinct and nonredundant in vivo functions of IFNAR on myeloid cells limit autoimmunity in the central nervous system. Immunity 28:675–686

99. Teijaro JR, Ng C, Lee AM, Sullivan BM, Sheehan KC et al (2013) Persistent LCMV infection is controlled by blockade of type I interferon signaling. Science 340:207–211
100. Wilson EB, Yamada DH, Elsaesser H, Herskovitz J, Deng J et al (2013) Blockade of chronic type I interferon signaling to control persistent LCMV infection. Science 340:202–207
101. Gonzalez-Navajas JM, Lee J, David M, Raz E (2012) Immunomodulatory functions of type I interferons. Nat Rev Immunol 12:125–135
102. Kovarik P, Sauer I, Schaljo B (2007) Molecular mechanisms of the anti-inflammatory functions of interferons. Immunobiology 212:895–901

Activation of Type I IFN Signaling by *Staphylococcus aureus*

Dane Parker

Introduction

Staphylococcus aureus is a Gram positive pathogen that is a significant cause of skin and soft tissue infections as well as pneumonia. *S. aureus* is the most common pathogen associated with skin and soft tissue infection resulting in over 11 million ambulatory care visits in the USA alone [1]. The emergence of antibiotic resistant strains of *S. aureus* is an increasing problem for the treatment of patients. The rise of methicillin resistant *S. aureus* (MRSA) is of particular concern. Methicillin resistance is encoded by the *mec* element and MRSA strains such as the USA300 isolates from the USA are highly prevalent within the community [2–4]. In the context of skin infections, USA300 strains possess the arginine catabolic mobile element (ACME) [5] that allows it to thrive in the acidic environment of the skin.

Lung infections cause more disease burden than cancer or HIV, with pneumonia leading to millions of deaths across the globe annually [6–8]. Pneumonia is also of increasing concern with an aging population and is the leading cause of hospitalization (disability-adjusted life-years lost) [9]. A large proportion of the population (30 %) [10] are asymptomatic carriers of *S. aureus* and evidence suggests that carriage increases the risk of infection. Chronic carriers have higher rates of infection and the strains isolated from infection sites are usually the same as those colonizing the nose [11–14]. There is a direct correlation between reducing disease and decreasing colonization and as such "de-colonization" of patients prior to surgery is common place and also provides economic savings [15–17].

D. Parker (✉)
Department of Pediatrics, Columbia University, New York, NY 10032, USA
e-mail: dp2375@cumc.columbia.edu

© Springer International Publishing Switzerland 2014

D. Parker (ed.), *Bacterial Activation of Type I Interferons*,
DOI 10.1007/978-3-319-09498-4_5

A vaccine does not exist for the prevention of *S. aureus* infection and while antibiotics are available to treat infection morbidity and mortality is still high [8]. Increasing evidence suggests that much of the pathology associated with *S. aureus* infection relates to the intensity of the host response and interventions that aim to subdue this response might prove beneficial in treatment [18, 19]. One such host pathway that has been examined is the type I interferon (IFN) pathway.

Co-infection Between *S. aureus* and Influenza

S. aureus is a major complication of influenza infection. Secondary pneumonia subsequent to influenza virus infection is a major cause of morbidity and mortality, with *S. aureus* being one of the major causative agents [7, 20, 21]. Analysis of bodies that contracted and subsequently died from the influenza pandemics of 1918 and 1957 show a high proportion of bacterial co-infection [22, 23]. Type I IFN signaling is known to be highly induced by influenza virus and activation of type I IFNs is important for host protection [24, 25]. Several studies have attempted to model influenza and *S. aureus* co-infection to address the role of type I IFNs in the increased susceptibility to *S. aureus* infection post influenza insult.

Models of *S. aureus* superinfection mimic the etiology of human infection. Mice are infected with influenza virus several days prior to bacterial insult. In this murine superinfection model mice that receive influenza virus fair worse than mice infected with bacteria alone, regardless of the route of infection (intranasal, intratracheal, or intravenous) [26, 27]. Superinfected mice exhibit increased pathology, with higher infiltrates, areas of necrosis, and score higher on outcomes of lung injury [26, 27]. In many cases this superinfection leads to increased mortality and higher bacterial loads in both the airway and lung tissue [27]. There are varying reports regarding a link between specific cell types contributing to the poorer outcome during superinfection. Lee et al. [27] observed increased neutrophil recruitment during superinfection with decreased levels of macrophages and dendritic cells. Kudva et al. [28] also observed increased neutrophils but also saw increased macrophages, while depletion of various cells types had no effect on bacterial counts or weight loss [26]. Studies by Kudva et al. [28] identified a link between type I IFN signaling and the enhanced susceptibility to bacterial infection.

Kudva et al. [28] observed that upon infection with *S. aureus* there were increased Th17 cells. They further showed that the Th17 cytokines such as IL-17 and IL-22 were involved in bacterial clearance. Mice unable to respond or produce IL-17 and IL-22 were impaired in their ability to clear *S. aureus;* however influenza infection led to decreased production of the Th17 cytokines IL-22 and IL-23 as well as gamma delta T cells that produce IL-17. IL-17-producing gamma delta T cells are known to be suppressed by type I IFNs [29]. Kudva et al. then showed that the influenza induced reduction in IL-17 was dependent on type I IFN signaling and thus were able to demonstrate a link between influenza infection, type I

IFN signaling and the ability to clear *S. aureus* from the lung [28]. The ability of type I IFN signaling to enhance susceptibility to infection has been further shown by pre-treating mice with the TLR3 ligand poly(I:C) followed by intratracheal administration of *S. aureus*. Poly(I:C) treated mice had increased bacterial loads and this effect was ameliorated in an *Ifnar*$^{-/-}$ background [30]. Likewise when *S. aureus* is given to mice, the type I IFNs generated lead to protection against viral infections [31].

The negative impact of type I IFN signaling on superinfection has also been studied in other models of infection. Prior infection of mice with lymphocytic choriomeningitis virus (LCMV) that stimulates a strong type I IFN response, followed by intravenous inoculation of *S. aureus* results in significant increases in bacterial loads in the lung and kidney [32]. In the absence of IFNAR, bacterial densities are equivalent to naïve mice not exposed to LCMV.

Role of Type I IFN Signaling in *S. aureus* Infection

The role of type I IFN signaling in primary infections with *S. aureus* has been examined in murine models. In models of primary respiratory infection with *S. aureus*, type I IFN signaling appears to play a negative role in outcome. *Ifnar*$^{-/-}$ mice have reduced mortality to *S. aureus* compared to wild-type mice. At lower, non-lethal doses of bacteria to examine the cellular response, a reduction in TNF is observed in the airway, while increases have been observed in serum [33]. The correlation between improved outcome and a reduction in TNF is consistent with previous studies [34]. Bacterial burden in *Ifnar*$^{-/-}$ mice varies between studies of pulmonary infection, from minor to a 20-fold reduction [33, 35]. It is possible that type I IFN signaling does not participate directly in bacterial clearance, but its influence on proinflammatory signaling does alter the eventual outcome of infection. This influence was evident when a strain of *S. aureus* that induces high levels of *Ifnb* was examined. *Ifnar*$^{-/-}$ mice had improved pulmonary pathology in response to *S. aureus* strain 502A (Fig. 1), reduced consolidation and improved alveolar architecture as well as reduction in proinflammatory cytokines such as KC and IL-1β [36]. In a model of skin infection it was observed that in contrast to the lung, induction of *Ifnb* is important in controlling the infection [37]. The addition of IFN-β to mice in a subcutaneous model of infection led to improved clearance of *S. aureus* and reduced lesion sizes, indicating differing roles for type I IFNs depending upon the site of infection.

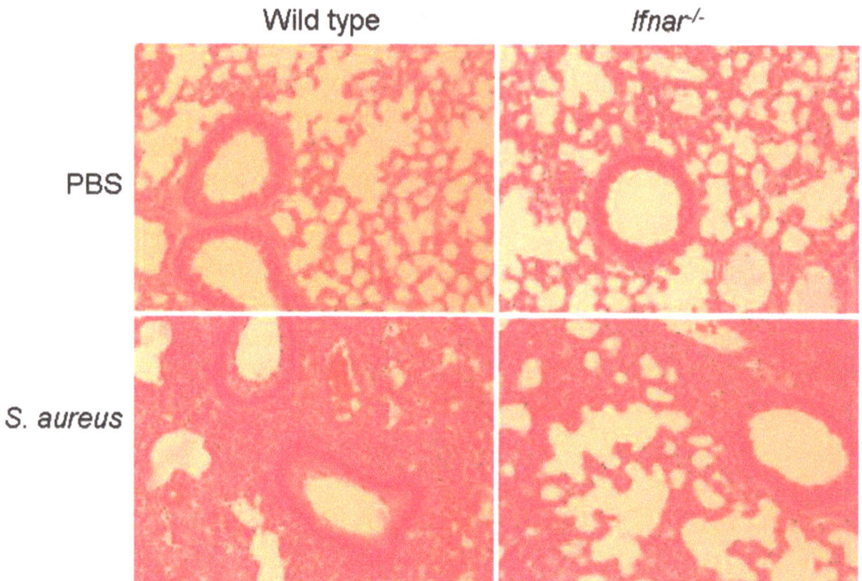

Fig. 1 Role of type I IFN in pulmonary pathology to *S. aureus* infection. Normal (C57Bl/6J) inbred mice and *Ifnar*$^{-/-}$ mice were infected with 10^7 colony forming units of *S. aureus* for 24 h. Lungs were fixed in 4 % paraformaldehdye before sectioning and staining with H&E. *Source*: Dane Parker and Alice Prince, Columbia University USA

Activation of Type I IFN Signaling by *S. aureus*

Given the importance of type I IFN signaling in *S. aureus* infection, the mechanisms for how it induces the interferon pathway have been examined in several cell types, including epithelial and dendritic cells. In airway epithelial cells [33] *S. aureus* USA300 rapidly activates the type I IFN pathway within 2 h. At this point phosphorylation of STAT1 and STAT2 is evident as well as induction of *Ifnb*, *Mx1*, *Lif*, and *Il6*. Although not proven in vivo, in vitro it is suggested that protein A of *S. aureus* activates type I IFN signaling. The most compelling data is that the variable region of protein A transfected into epithelial cells can induce *Ifnb* expression while those with the control vector do not [33].

The ability of *S. aureus* to activate type I IFN signaling in phagocytic cells has been well documented [31, 35, 38, 39]. In dendritic cells (DCs) it has been observed that several of the major factors of USA300 *S. aureus* are not involved in type I IFN activation, such as α-toxin, Panton Valentine toxin, and protein A. Relatively high levels of *Ifnb* induction are observed upon stimulation with *S. aureus*. As observed in epithelial cells [33], upon stimulation by *S. aureus*, DCs phosphorylate STAT1 and lead to induction of genes such as *Ifnb*, *Mx1*, and *Cxcl10* within hours. The induction of these genes is dependent upon autocrine signaling as cells lacking IFNAR show reduced gene induction [35]. The induction of type I IFN by *S. aureus*

USA300 in DCs is not dependent on surface TLR expression; however, DCs do require uptake of *S. aureus* to signal type I IFN. Inhibition of phagocytosis with various inhibitors negates *Ifnb* induction, as does inhibition of endosomal acidification with chloroquine [35, 37, 40]. Inhibition by chloroquine suggests the involvement of an endosomal receptor and it has been shown that TLR9 and MyD88 are required to induce type I IFN signaling. Additional adapters such as TRIF or other endosomal receptors such as TLR7 are not involved. Consistent with the role of TLR9, IRF1 was also found to be involved in the *S. aureus* induced signaling cascade [41]. Although not examined by the Parker et al. study [35], previous work has identified that lipoteichoic acid of *S. aureus* can increase IFN-α via IRF2 and IRF1 [42]. The role of *S. aureus* DNA in signaling type I IFN via TLR9 was confirmed using bacterial lysates that had been treated with nucleases, with DNase causing a significant decrease in type I IFN signaling (Fig. 2) [35]. It is likely then that this sensing by TLR9 is part of normal cell processing of bacteria and not an active response induced by *S. aureus*.

At the gene level, $Tlr9^{-/-}$ DCs did not show changes in proinflammatory gene transcription, the exception being *Cxcl10*. However, at the protein level a reduction in TNF was observed. The reduction in TNF by $Tlr9^{-/-}$ DCs was not as great as that observed in $MyD88^{-/-}$ DCs in response to *S. aureus*. The importance of TLR9 has been tested in a murine model of acute pneumonia. $Tlr9^{-/-}$ mice show an improved response compared to wild-type mice when infected with *S. aureus* USA300.

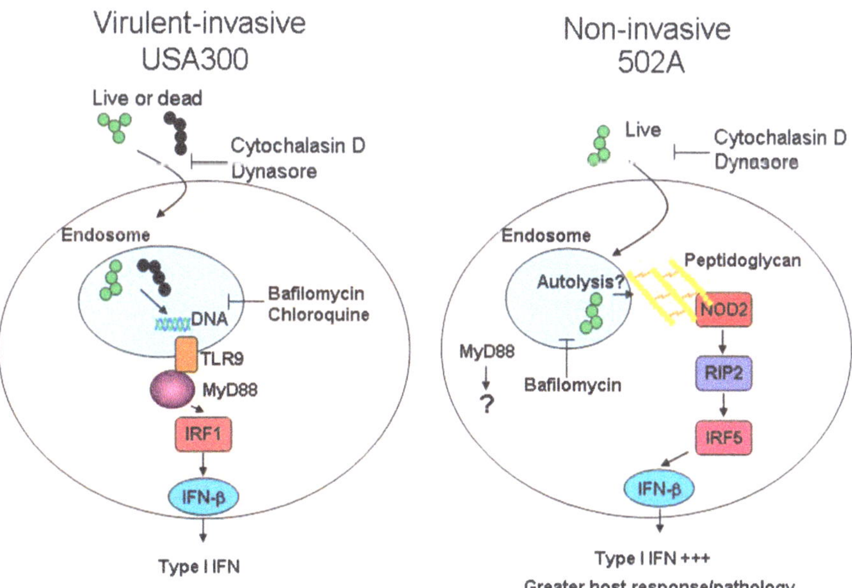

Fig. 2 Activation of type I IFN signaling by *S. aureus*. Receptors and pathways involved in activation of type I IFN signaling by *S. aureus* strains in dendritic cells

Mice lacking TLR9 had improved clearance of bacteria from the airway and at higher inoculums had reduced rates of mortality. Minimal cytokine differences are observed between wild-type of *Tlr9*$^{-/-}$ mice, with the major difference being a reduction in TNF. This result is not surprising given the known negative role TNF plays in *S. aureus* pneumonia [34].

Recent work has indicated that TLR9 is not the only receptor involved in type I IFN activation by *S. aureus* and there exists variation within the species. This observation was made using the *S. aureus* strain 502A. 502A is an important clinical strain used in the 1960s for bacterial interference studies, whereby neonates were inoculated with 502A to protect them from infection by the circulating invasive strain during that time period [43]. This ability to colonize and not invade by 502A has been replicated in several skin and airway epithelial cells lines [36]. Comparison of the USA300 and 502A genomes revealed that 502A encodes a similar virulence factor repertoire to USA300. Studies undertaken to determine the host response to these clinically disparate strains identified a significant difference in the ability of 502A to induce type I IFN signaling. 502A induced significantly higher levels of *Ifnb* compared to USA300 in various cell types [36]. In phagocytic cells *Ifnb* was the only gene upregulated, with other proinflammatory genes unchanged. This observation suggested type I IFN signaling may be important in the ability of 502A to prevent infection from invasive isolates.

The mechanism of type I IFN induction by 502A differs to that of USA300. Unlike USA300, live 502A is required to induce an interferon response. Heat-killed organisms induce significantly less *Ifnb*, to levels analogous of USA300. In DC, uptake is still required but chloroquine does not inhibit induction of *Ifnb*, indicating that 502A does not signal through TLR9 [36]. Screening of several TLR and adapter proteins identified a unique pathway to IFN activation. *S. aureus* 502A signals via the peptidoglycan receptor NOD2. Consistent with a role for NOD2, its downstream kinase RIP2 and IRF5 were also involved (Fig. 2) [36]. Several potential interferon agonists (i.e., DNA, RNA, cell wall extracts) from 502A and USA300 were compared, but all gave similar levels of induction, including peptidoglycan, indicating that it was not a structural difference in peptidoglycan causing the enhanced response by 502A, but potentially differential regulation and/or release.

One mechanism for the enhanced interferon response by 502A is autolysis. Autolysis is a natural bacterial process that leads to cellular lysis and subsequent release of intracellular products and cell wall fragments. Proteomic analysis of secreted proteins by USA300 and 502A identified that 502A secreted more of the major staphylococcal autolysin, Atl. 502A displays enhanced autolysis in triton X-100 autolysis assays as well as an increased growth rate. Consistent with a role for peptidoglycan and autolysis in type I IFN activation by 502A, 502A is also more susceptible to lysostaphin [36]. Lysostaphin, produced by other species of staphylococci, is an endopeptidase that cleaves cross-links in the peptidoglycan of staphylococci. These observations are partially consistent with other reports that *S. aureus* is a naturally low inducer of type I IFNs and this low induction is due to lysozyme resistance [37]. While 502A is equally resistant to lysozyme, this is further evidence that the synthesis and turnover of the cell wall are likely to play a role in the ability of *S. aureus* to activate this host pathway.

Conclusion

S. aureus is an important bacterial pathogen that still causes significant morbidity and mortality with the advent of antibiotics. The ability of this pathogen to activate the type I IFN pathway is clearly important in the context of pulmonary infection, be it primary or secondary pneumonia post influenza, in addition to skin/soft tissue infections. The ability of *S. aureus* to activate this pathway does not appear to be a single process, with multiple receptors involved that varies between strains, as well as the involvement of several different cell types. Future studies need to be focused on determining the host factors that contribute to the immunopathology of interferons and their ability to influence the outcome of infection, as well as their role in pathogenesis at different bodily sites. These studies will open novel avenues in immunomodulary therapy.

References

1. McCaig LF, McDonald LC, Mandal S, Jernigan DB (2006) *Staphylococcus aureus*-associated skin and soft tissue infections in ambulatory care. Emerg Infect Dis 12:1715–1723
2. Diep BA, Gill SR, Chang RF, Phan TH, Chen JH, Davidson MG, Lin F, Lin J, Carleton HA, Mongodin EF, Sensabaugh GF, Perdreau-Remington F (2006) Complete genome sequence of USA300, an epidemic clone of community-acquired meticillin-resistant *Staphylococcus aureus*. Lancet 367:731–739
3. McDougal LK, Steward CD, Killgore GE, Chaitram JM, McAllister SK, Tenover FC (2003) Pulsed-field gel electrophoresis typing of oxacillin-resistant *Staphylococcus aureus* isolates from the United States: establishing a national database. J Clin Microbiol 41:5113–5120
4. Uhlemann AC, Knox J, Miller M, Hafer C, Vasquez G, Ryan M, Vavagiakis P, Shi Q, Lowy FD (2011) The environment as an unrecognized reservoir for community-associated methicillin resistant *Staphylococcus aureus* USA300: a case-control study. PLoS One 6:e22407
5. Thurlow LR, Joshi GS, Clark JR, Spontak JS, Neely CJ, Maile R, Richardson AR (2013) Functional modularity of the arginine catabolic mobile element contributes to the success of USA300 methicillin-resistant *Staphylococcus aureus*. Cell Host Microbe 13:100–107
6. Taubes G (2008) The bacteria fight back. Science 321:356–361
7. Kollef MH, Micek ST (2005) *Staphylococcus aureus* pneumonia: a "superbug" infection in community and hospital settings. Chest 128:1093–1097
8. Klevens RM, Morrison MA, Nadle J, Petit S, Gershman K, Ray S, Harrison LH, Lynfield R, Dumyati G, Townes JM, Craig AS, Zell ER, Fosheim GE, McDougal LK, Carey RB, Fridkin SK; Active Bacterial Core surveillance (ABCs) MRSA Investigators (2007) Invasive methicillin-resistant *Staphylococcus aureus* infections in the United States. JAMA 298: 1763–1771
9. Mizgerd JP (2012) Respiratory infection and the impact of pulmonary immunity on lung health and disease. Am J Respir Crit Care Med 186:824–829
10. Johannessen M, Sollid JE, Hanssen AM (2012) Host- and microbe determinants that may influence the success of *S. aureus* colonization. Front Cell Infect Microbiol 2:56
11. von Eiff C, Becker K, Machka K, Stammer H, Peters G (2001) Nasal carriage as a source of *Staphylococcus aureus* bacteremia. Study Group. N Engl J Med 344:11–16
12. Vuononvirta J, Toivonen L, Grondahl-Yli-Hannuksela K, Barkoff AM, Lindholm L, Mertsola J, Peltola V, He Q (2011) Nasopharyngeal bacterial colonization and gene polymorphisms of mannose-binding lectin and toll-like receptors 2 and 4 in infants. PLoS One 6:e26198

13. Wertheim HF, Vos MC, Ott A, van Belkum A, Voss A, Kluytmans JA, van Keulen PH, Vandenbroucke-Grauls CM, Meester MH, Verbrugh HA (2004) Risk and outcome of nosocomial *Staphylococcus aureus* bacteraemia in nasal carriers versus non-carriers. Lancet 364: 703–705
14. Yu VL, Goetz A, Wagener M, Smith PB, Rihs JD, Hanchett J, Zuravleff JJ (1986) *Staphylococcus aureus* nasal carriage and infection in patients on hemodialysis. Efficacy of antibiotic prophylaxis. N Engl J Med 315:91–96
15. Diller R, Sonntag AK, Mellmann A, Grevener K, Senninger N, Kipp F, Friedrich AW (2008) Evidence for cost reduction based on pre-admission MRSA screening in general surgery. Int J Hyg Environ Health 211:205–212
16. Lederer SR, Riedelsdorf G, Schiffl H (2007) Nasal carriage of meticillin resistant *Staphylococcus aureus*: the prevalence, patients at risk and the effect of elimination on outcomes among outclinic haemodialysis patients. Eur J Med Res 12:284–288
17. Kluytmans J (1998) Reduction of surgical site infections in major surgery by elimination of nasal carriage of *Staphylococcus aureus*. J Hosp Infect 40(Suppl B):S25–S29
18. Parker D, Prince A (2012) Immunopathogenesis of *Staphylococcus aureus* pulmonary infection. Semin Immunopathol 34:281–297
19. DeLeo FR, Otto M, Kreiswirth BN, Chambers HF (2010) Community-associated meticillin-resistant *Staphylococcus aureus*. Lancet 375:1557–1568
20. Hageman JC, Uyeki TM, Francis JS, Jernigan DB, Wheeler JG, Bridges CB, Barenkamp SJ, Sievert DM, Srinivasan A, Doherty MC, McDougal LK, Killgore GE, Lopatin UA, Coffman R, MacDonald JK, McAllister SK, Fosheim GE, Patel JB, McDonald LC (2006) Severe community-acquired pneumonia due to *Staphylococcus aureus*, 2003–04 influenza season. Emerg Infect Dis 12:894–899
21. Dawood FS, Chaves SS, Perez A, Reingold A, Meek J, Farley MM, Ryan P, Lynfield R, Morin C, Baumbach J, Bennett NM, Zansky S, Thomas A, Lindegren ML, Schaffner W, Finelli L, Emerging Infections Program Network (2014) Complications and associated bacterial co-infections among children hospitalized with seasonal or pandemic influenza, United States, 2003–2010. J Infect Dis 209:686–694
22. Morens DM, Taubenberger JK, Fauci AS (2008) Predominant role of bacterial pneumonia as a cause of death in pandemic influenza: implications for pandemic influenza preparedness. J Infect Dis 198:962–970
23. Louria DB, Blumenfeld HL, Ellis JT, Kilbourne ED, Rogers DE (1959) Studies on influenza in the pandemic of 1957–1958. II. Pulmonary complications of influenza. J Clin Invest 38: 213–265
24. Muller U, Steinhoff U, Reis LF, Hemmi S, Pavlovic J, Zinkernagel RM, Aguet M (1994) Functional role of type I and type II interferons in antiviral defense. Science 264:1918–1921
25. Li W, Moltedo B, Moran TM (2012) Type I interferon induction during influenza virus infection increases susceptibility to secondary *Streptococcus pneumoniae* infection by negative regulation of gammadelta T cells. J Virol 86:12304–12312
26. Iverson AR, Boyd KL, McAuley JL, Plano LR, Hart ME, McCullers JA (2011) Influenza virus primes mice for pneumonia from *Staphylococcus aureus*. J Infect Dis 203:880–888
27. Lee MH, Arrecubieta C, Martin FJ, Prince A, Borczuk AC, Lowy FD (2010) A postinfluenza model of *Staphylococcus aureus* pneumonia. J Infect Dis 201:508–515
28. Kudva A, Scheller EV, Robinson KM, Crowe CR, Choi SM, Slight SR, Khader SA, Dubin PJ, Enelow RI, Kolls JK, Alcorn JF (2011) Influenza A inhibits Th17-mediated host defense against bacterial pneumonia in mice. J Immunol 186:1666–1674
29. Henry T, Kirimanjeswara GS, Ruby T, Jones JW, Peng K, Perret M, Ho L, Sauer JD, Iwakura Y, Metzger DW, Monack DM (2010) Type I IFN signaling constrains IL-17A/F secretion by gammadelta T cells during bacterial infections. J Immunol 184:3755–3767
30. Tian X, Xu F, Lung WY, Meyerson C, Ghaffari AA, Cheng G, Deng JC (2012) Poly I:C enhances susceptibility to secondary pulmonary infections by gram-positive bacteria. PLoS One 7:e41879

31. Perkins DJ, Polumuri SK, Pennini ME, Lai W, Xie P, Vogel SN (2013) Reprogramming of murine macrophages through TLR2 confers viral resistance via TRAF3-mediated, enhanced interferon production. PLoS Pathog 9:e1003479
32. Navarini AA, Recher M, Lang KS, Georgiev P, Meury S, Bergthaler A, Flatz L, Bille J, Landmann R, Odermatt B, Hengartner H, Zinkernagel RM (2006) Increased susceptibility to bacterial superinfection as a consequence of innate antiviral responses. Proc Natl Acad Sci U S A 103:15535–15539
33. Martin FJ, Gomez MI, Wetzel DM, Memmi G, O'Seaghdha M, Soong G, Schindler C, Prince A (2009) *Staphylococcus aureus* activates type I IFN signaling in mice and humans through the Xr repeated sequences of protein A. J Clin Invest 119:1931–1939
34. Gomez MI, Lee A, Reddy B, Muir A, Soong G, Pitt A, Cheung A, Prince A (2004) *Staphylococcus aureus* protein A induces airway epithelial inflammatory responses by activating TNFR1. Nat Med 10:842–848
35. Parker D, Prince A (2012) *Staphylococcus aureus* induces type I IFN signaling in dendritic cells via TLR9. J Immunol 189:4040–4046
36. Parker D, Planet PJ, Soong G, Narechania A, Prince A (2014) Induction of type I interferon signaling determines the relative pathogenicity of *Staphylococcus aureus* strains. PLoS Pathog 10:e1003951
37. Kaplan A, Ma J, Kyme P, Wolf AJ, Becker CA, Tseng CW, Liu GY, Underhill DM (2012) Failure to induce IFN-beta production during *Staphylococcus aureus* infection contributes to pathogenicity. J Immunol 189:4537–4545
38. Svensson H, Cederblad B, Lindahl M, Alm G (1996) Stimulation of natural interferon-alpha/beta-producing cells by *Staphylococcus aureus*. J Interferon Cytokine Res 16:7–16
39. Michea P, Vargas P, Donnadieu MH, Rosemblatt M, Bono MR, Dumenil G, Soumelis V (2013) Epithelial control of the human pDC response to extracellular bacteria. Eur J Immunol 43: 1264–1273
40. Ip WK, Sokolovska A, Charriere GM, Boyer L, Dejardin S, Cappillino MP, Yantosca LM, Takahashi K, Moore KJ, Lacy-Hulbert A, Stuart LM (2010) Phagocytosis and phagosome acidification are required for pathogen processing and MyD88-dependent responses to *Staphylococcus aureus*. J Immunol 184:7071–7081
41. Schmitz F, Heit A, Guggemoos S, Krug A, Mages J, Schiemann M, Adler H, Drexler I, Haas T, Lang R, Wagner H (2007) Interferon-regulatory-factor 1 controls Toll-like receptor 9-mediated IFN-beta production in myeloid dendritic cells. Eur J Immunol 37:315–327
42. Liljeroos M, Vuolteenaho R, Rounioja S, Henriques-Normark B, Hallman M, Ojaniemi M (2008) Bacterial ligand of TLR2 signals Stat activation via induction of IRF1/2 and interferon-alpha production. Cell Signal 20:1873–1881
43. Shinefield HR, Ribble JC, Eichenwald HF, Boris M, Sutherland JM (1963) Bacterial interference: its effect on nursery-acquired infection with *Staphylococcus aureus*. V. An analysis and interpretation. Am J Dis Child 105:683–688

Contribution of Interferon Signaling to Host Defense Against *Pseudomonas aeruginosa*

Taylor S. Cohen and Alice Prince

Introduction

Pseudomonas aeruginosa, an opportunistic bacterial pathogen, is not normally a component of the airway flora but is ubiquitous in the environment and especially common in health care-associated facilities [1, 2]. Aspiration or contamination of the airways with *P. aeruginosa* is an infrequent cause of pneumonia in a normal host, but is a common pathogen in immunocompromised and mechanically ventilated patients [1–4]. Clearance of *P. aeruginosa* from the host depends on recognition of the bacteria by the innate immune receptors, recruitment of phagocytic cells to the site of infection, and anti-inflammatory signaling to minimize tissue damage. This chapter will focus on the role of type I interferon (IFN) in the host response to *P. aeruginosa*, including how this pathway is activated, what signaling occurs downstream of type I IFNs, how they contribute to the host response, and specific genetic disorders that influence induction of type I IFNs.

Activation of Type I Interferon Signaling

P. aeruginosa presents an array of pathogen-associated molecular patterns (PAMPs) such as lipopolysaccharide (LPS), lipoproteins, and flagellin that initiate host signaling. These PAMPs are recognized by an array of receptors, available at the cell surface or internally within endosomes or free in the cytoplasm. Toll-like receptors (TLRs) such as TLR2, TLR4, and TLR5 are expressed on the cell surface and signal

T.S. Cohen • A. Prince (✉)
Department of Pediatrics, Columbia University, 650 W 168th Street, New York, NY 10032, USA
e-mail: asp7@cumc.columbia.edu

© Springer International Publishing Switzerland 2014

D. Parker (ed.), *Bacterial Activation of Type I Interferons*,
DOI 10.1007/978-3-319-09498-4_6

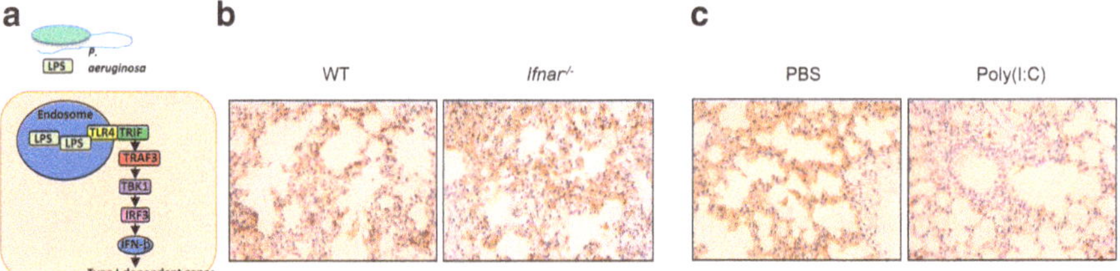

Fig. 1 Type I IFN signaling pathway contributes to *P. aeruginosa* clearance. (**a**) TLR4-TRIF signaling pathway. (**b**) TUNEL stained lung sections from PAK-infected WT or *Ifnar*$^{-/-}$ mice. (**c**) TUNEL stained lung sections from PAK-infected WT mice pretreated with PBS or Poly(I:C)

through the adaptor protein MyD88 to activate innate immune signaling. Recognition of PAMPs by intracellular receptors is generally required to initiate type I IFN signaling, requiring the internalization of specific bacterial PAMPs such LPS and DNA [5, 6]. TLR4 is uniquely expressed on the cell surface and within endosomes where in conjunction with the adaptor TRIF it can activate type I IFN signaling via IRF3 (Fig. 1a) [7]. Of note, the majority, almost 75 %, of TLR4-dependent genes are controlled in a MyD88-independent manner [8]. As an airway pathogen, *P. aeruginosa* sheds LPS into the airway lumen enabling uptake by epithelial as well as immune cells activating endosomal TLR4/TRIF signaling [9, 10]. In the context of *P. aeruginosa*, TLR4 has been shown to be the primary receptor through which *P. aeruginosa* activates production of type I IFN [11].

Contribution of Type I Signaling to Host Defense Against *P. aeruginosa*

P. aeruginosa is sensed by both airway epithelial cells and immune cells, activating primarily proinflammatory chemokines and cytokine expression through host recognition of bacterial PAMPs. In addition to common proinflammatory cytokines such as CXCL8 and TNF, type I IFN is also produced by epithelial cells and immune cells in response to the pathogen. Signaling through the ubiquitously expressed type I IFN receptor IFNAR and JAK/STAT kinases, type I IFN activates greater than 300 downstream genes that contribute to the host response. The vast array of downstream genes and the global expression of IFNAR suggest that type I IFN participate in host defense in a tissue and "model" dependent fashion.

The role of type I IFN in host defense against *P. aeruginosa* has been examined in the context of both respiratory infection and sepsis. Some studies suggest a beneficial contribution of type I IFN signaling to the host response to *P. aeruginosa*, while others demonstrate little to no effect of inhibiting or activating this pathway. Initial studies in the lung were conducted using TRIF-deficient mice, which lack a key adapter protein linking TLR4 and type I IFN production, and demonstrated a requirement for TRIF-dependent signaling for the clearance of *P. aeruginosa* from the lung [12]. Macrophages lacking TRIF produced lower amounts of CCL5

(RANTES), TNF, and KC compared to wild-type macrophages in response to *P. aeruginosa*. Similarly, lower levels of these cytokines were found in the lung of TRIF null mice following infection. TRIF induces both NF-κB and IRF3-dependent signaling, as evidenced by significant reductions of KC and CCL5 in the TRIF knockout. A second study by the same group looked more specifically at the IRF3-IFN arm of the TRIF signaling pathway using mice lacking IRF3 [13]. Clearance of *P. aeruginosa* was significantly impaired in IRF3 null mice, as was production of type I IFN and RANTES. Expression of NF-κB-dependent genes such as KC or TNF was not affected by the knockout of IRF3. To demonstrate more clearly that type I IFN is directly responsible for the clearance of *P. aeruginosa* the authors attempted to reconstitute the system with IFN-β but were unsuccessful possibly for technical reasons. A more direct assessment of the role of type I IFN was done using mice lacking the type I IFN receptor, IFNAR; and no difference in acute (18 h) clearance of *P. aeruginosa* from the lungs of IFNAR compared to WT mice or lung damage was observed (Fig. 1b) [14].

Beneficial effects of type I IFN signaling have been documented. Stimulation of the TLR3-TRIF-IRF3 pathway with Poly(I:C) 24 h prior to infection with *P. aeruginosa* improved clearance from the lung, while reducing expression of IL-1β and IL-18 in the airway and lung pathology (Fig. 1c). Expression of other pro-inflammatory cytokines such as KC and TNF were not affected. In vitro, IFN-β pretreatment of alveolar macrophages decreases IL-1β production in response to *P. aeruginosa* [15]. These results, combined with elevated IL-1β levels observed in the airways of IFNAR null mice compared to WT mice at baseline, suggest that type I IFN acts to limit inflammatory cytokine production over the course of *P. aeruginosa* infection and perhaps reduce the inflammatory tone in the lung. These data suggest a potential therapeutic use for type I IFN therapies in the context of *P. aeruginosa* and other inflammatory pneumonias.

The role of type I IFN has been less well studied in the context of other models of *P. aeruginosa* infection: results from two sepsis models came to opposite conclusions on the role of type I IFN. In vivo models of sepsis suggest a role for type I IFN and specifically the downstream effector CXCL10 in the response to polymicrobial sepsis induced in the cecal ligation and puncture model. Using the IFNAR knockout mice there was a requirement for type I IFN for survival following a polymicrobial insult, cecal ligation, and puncture, yet a detrimental effect of type I IFN during LPS endotoxicosis [16]. In the context of a polymicrobial insult, signaling through IFNAR expressed on hematopoetic cells regulated a select set of cytokines including CXCL10. Addition of CXCL10 to IFNAR mice improved survival do in part to increased phagocytic capacity of neutrophils. Enhanced neutrophil survival during LPS endotoxicosis, a less physiologic model in which phagocytosis of bacteria is not required, could contribute to increased tissue inflammation and possibly explain the contrasting roles of type I IFN in this study.

A study by Dejager et al. [17] also examined the response of IFNAR null mice to polymicrobial sepsis, and found improved survival of IFNAR null mice compared to wild-type mice in the context of a polymicrobial insult, LPS endotoxicosis, or heat killed *P. aeruginosa*. Importantly, this group demonstrated that neutralizing the type I IFN pathway in wild-type mice with antibody against IFNAR was protective.

Some key differences such as strain of mouse, severity of the insult, and the use of antibiotics exist and confound the interpretation of the two studies. This group administered antibiotics to the mice following cecal ligation and puncture to mimic clinical treatment, which makes comparison of this model to that of Kelly-Scumpia difficult. Killing of bacteria by antibiotics makes phagocytic killing of bacteria by immune cells unnecessary, skewing the model such that inflammation caused by immune cells is more damaging than protective. In fact, protective effects of type I IFN inhibition were primarily observed with high dose antibiotic treatment, suggesting a proinflammatory role for type I IFN during sterile inflammation. Similar proinflammatory action of type I IFN has been observed in models of systemic TNF-induced inflammation [18].

Linking Type I Interferon to Cystic Fibrosis

Patients with cystic fibrosis (CF) become colonized with *P. aeruginosa* leading to eventual respiratory failure due to the chronic inflammatory state of the airways associated with the infection [19]. The literature suggests that initial colonization of the CF lung by *P. aeruginosa* is due to an epithelial specific defect in the ability to induce type I IFN signaling in response to TLR4 stimulation [11]. As a consequence, epithelial participation in the innate immunity is compromised.

Epithelial production of type I IFN results in activation of immune cell populations, an event observed both in vitro in a human model system in which CF epithelial cell supernatants were unable to activate peripheral blood dendritic cells and in vivo in mice receiving poly(I:C) prior to infection [11]. A similar observation was reported in mice lacking IRF3, a major component of the TLR4-IFN, were limited in their ability to recruit macrophages and neutrophils to the airway in response to *P. aeruginosa* [13]. Macrophage polarization and cytokine production are regulated by IFN signaling, and in the CF lung Th2 cytokines are predominant [20]. Type I IFN drives M1 macrophages that produce some of the Th1 family of cytokines; therefore reduced type I IFN in the CF airway could underlie the observed cytokine imbalance [21]. While the link between epithelial production of type I IFN and activation of immune cells in the CF lung has yet to be made in patients, these data suggest that a defect in the CF epithelial cell related to the induction of type I IFN could contribute to respiratory colonization by *P. aeruginosa* and lung inflammation.

Contribution of Other Interferons

P. aeruginosa induces host production of type II and type III IFN in addition to type I IFN. Type II IFN, or IFN gamma, has been linked to tissue inflammation in a model of systemic *P. aeruginosa* infection following cecal ligation and puncture [22]. The IFN gamma knockout mouse was observed to have reduced IL-6 and elevated IL-10 following *P. aeruginosa* infection, and were unable to eliminate the bacteria, do in part to a requirement for IFN gamma mediated macrophage killing of *P. aeruginosa* [23]. Reduced signaling induction in the knockout mouse could also be attributed to reduced MD2 expression, a key chaperone protein for LPS-TLR4 interaction that is regulated by type II IFN [24]. Addition of exogenous type II IFN to WT mice promoted inflammatory cytokine production while not affecting bacterial clearance, suggesting that IFN gamma's contribution to inflammation is greater than its role in bacterial clearance.

Type III IFN is the least studied of the IFNs, especially in the context of bacterial infection. Similar to type I IFN, type III IFN activates the same set of greater than 300 IFN-dependent genes downstream of JAK/STAT kinases [25–28]. *P. aeruginosa* activates type III IFN signaling during respiratory infection, and the secreted cytokine is sensed by respiratory epithelial cells [14]. As opposed to IFNAR null mice that clear *P. aeruginosa* in a manner comparable to wild-type mice, mice lacking the receptor for type III IFN, IL-28R, have improved clearance of the bacteria from the lung [14]. Improved clearance correlates with reduced expression of inflammatory cytokines in the airway and improved lung pathology. Unlike the ubiquitous expression of IFNAR, IL-28R expression is restricted to mucosal epithelial cells [29]. A more limited cellular response to type III IFN in the lung could explain the differing phenotypes observed in the type I and type III IFN-specific knockout mice, although additional research is required to elucidate the specific downstream mechanisms.

Conclusion

The opportunist human pathogen *P. aeruginosa* is a potent stimulator of IFN signaling in the host. Activating the TLR4-TRIF-IRF3-IFN pathway through release of LPS or DAI in response to bacterial DNA induces secretion of type I IFNs from innate immune cells and airway epithelial cells. Type I IFN, in the context of *P. aeruginosa* pneumonia, regulates production of inflammatory cytokines such as IL-1β and IL-18 limiting the degree of host pathology. A similar protective effect was observed in a sepsis model of infection, although these results have yet to be confirmed. Therefore in the context of *P. aeruginosa* infection it seems that induction of the type I IFN pathway plays a host protective role.

References

1. Richards MJ, Edwards JR, Culver DH, Gaynes RP (1999) Nosocomial infections in medical intensive care units in the United States. National Nosocomial Infections Surveillance System. Crit Care Med 27:887–892
2. Lynch JP (2001) Hospital-acquired pneumonia: risk factors, microbiology, and treatment. Chest 119:373S–384S

3. Rodríguez-Rojas A, Oliver A, Blázquez J (2012) Intrinsic and environmental mutagenesis drive diversification and persistence of *Pseudomonas aeruginosa* in chronic lung infections. J Infect Dis 205:121–127
4. Craven DE, Hjalmarson KI (2010) Ventilator associated tracheobronchitis and pneumonia: thinking outside the box. Clin Infect Dis 51:S59–S66
5. Kagan JC, Su T, Horng T et al (2008) TRAM couples endocytosis of Toll-like receptor 4 to the induction of interferon-beta. Nat Immunol 9:361–368
6. Parker D, Martin FJ, Soong G et al (2011) *Streptococcus pneumoniae* DNA initiates type I interferon signaling in the respiratory tract. MBio 2:e00016-11
7. Doyle S, Vaidya S, O'Connell R et al (2002) IRF3 mediates a TLR3/TLR4-specific antiviral gene program. Immunity 17:251–263
8. Björkbacka H, Fitzgerald KA, Huet F et al (2004) The induction of macrophage gene expression by LPS predominantly utilizes Myd88-independent signaling cascades. Physiol Genomics 19:319–330
9. Rowe DC, McGettrick AF, Latz E et al (2006) The myristoylation of TRIF-related adaptor molecule is essential for Toll-like receptor 4 signal transduction. Proc Natl Acad Sci U S A 103:6299–6304
10. Visintin A, Mazzoni A, Spitzer JA, Segal DM (2001) Secreted MD-2 is a large polymeric protein that efficiently confers lipopolysaccharide sensitivity to Toll-like receptor 4. Proc Natl Acad Sci U S A 98:12156–12161
11. Parker D, Cohen TS, Alhede M et al (2012) Induction of type I interferon signaling by *Pseudomonas aeruginosa* is diminished in cystic fibrosis epithelial cells. Am J Respir Cell Mol Biol 46:6–13
12. Power MR, Li B, Yamamoto M et al (2007) A role of Toll-IL-1 receptor domain-containing adaptor-inducing IFN-beta in the host response to *Pseudomonas aeruginosa* lung infection in mice. J Immunol 178:3170–3176
13. Carrigan SO, Junkins R, Yang YJ et al (2010) IFN regulatory factor 3 contributes to the host response during *Pseudomonas aeruginosa* lung infection in mice. J Immunol 185:3602–3609
14. Cohen TS, Prince AS (2013) Bacterial pathogens activate a common inflammatory pathway through IFNλ regulation of PDCD4. PLoS Pathog 9:e1003682
15. Cohen TS, Prince AS (2013) Activation of inflammasome signaling mediates pathology of acute *P. aeruginosa* pneumonia. J Clin Invest 123:1630–1637
16. Kelly-Scumpia KM, Scumpia PO, Delano MJ et al (2010) Type I interferon signaling in hematopoietic cells is required for survival in mouse polymicrobial sepsis by regulating CXCL10. J Exp Med 207:319–326
17. Dejager L, Vandevyver S, Ballegeer M et al (2013) Pharmacological inhibition of type I interferon signaling protects mice against lethal sepsis. J Infect Dis 1–32
18. Huys L, Van Hauwermeiren F, Dejager L et al (2009) Type I interferon drives tumor necrosis factor-induced lethal shock. J Exp Med 206:1873–1882
19. Cohen TS, Prince A (2012) Cystic fibrosis: a mucosal immunodeficiency syndrome. Nat Med 18:509–519
20. Tiringer K, Treis A, Fucik P et al (2013) A Th17- and Th2-skewed cytokine profile in cystic fibrosis lungs represents a potential risk factor for *Pseudomonas aeruginosa* infection. Am J Respir Crit Care Med 187:621–629
21. Porta C, Rimoldi M, Raes G et al (2009) Tolerance and M2 (alternative) macrophage polarization are related processes orchestrated by p50 nuclear factor kappaB. Proc Natl Acad Sci U S A 106:14978–14983
22. Murphey ED, Herndon DN, Sherwood ER (2004) Gamma interferon does not enhance clearance of *Pseudomonas aeruginosa* but does amplify a proinflammatory response in a murine model of postseptic immunosuppression. Infect Immun 72:6892–6901
23. Leid JG, Willson CJ, Shirtliff ME et al (2005) The exopolysaccharide alginate protects *Pseudomonas aeruginosa* biofilm bacteria from IFN-gamma-mediated macrophage killing. J Immunol 175:7512–7518

24. Roy S, Sun Y, Pearlman E (2011) Interferon-induced MD-2 protein expression and lipopolysaccharide (LPS) responsiveness in corneal epithelial cells is mediated by Janus tyrosine kinase-2 activation and direct binding of STAT1 protein to the MD-2 promoter. J Biol Chem 286:23753–23762
25. Kotenko SV, Gallagher G, Baurin VV et al (2003) IFN-lambdas mediate antiviral protection through a distinct class II cytokine receptor complex. Nat Immunol 4:69–77
26. Sheppard P, Kindsvogel W, Xu W et al (2003) IL-28, IL-29 and their class II cytokine receptor IL-28R. Nat Immunol 4:63–68
27. Meager A, Visvalingam K, Dilger P et al (2005) Biological activity of interleukins-28 and -29: comparison with type I interferons. Cytokine 31:109–118
28. Maher SG, Sheikh F, Scarzello AJ et al (2008) IFNalpha and IFNlambda differ in their antiproliferative effects and duration of JAK/STAT signaling activity. Cancer Biol Ther 7: 1109–1115
29. Sommereyns C, Paul S, Staeheli P, Michiels T (2008) IFN-lambda (IFN-lambda) is expressed in a tissue-dependent fashion and primarily acts on epithelial cells in vivo. PLoS Pathog 4:e1000017

The Detrimental Role of Type I Interferon Signaling During Infection with *Salmonella typhimurium*

Bojan Shutinoski and Subash Sad

TLR-Dependent Induction of Type I IFNs

During infectious disease a host can recognize pathogens by various receptors unified under the term pathogen recognition receptors (PRR). These receptors initiate signaling cascades to alert the host immune system of the imminent danger associated with the invading pathogen, which commits the immune system to first restraining and ultimately, clearing off the pathogen [1]. Such signaling results in activation of the innate immune response, which in turn leads to amplification of the adaptive branch of the immune system [2]. Among the best described PRRs is the family of the Toll-like receptors (TLRs). These receptors are expressed by most if not all host cells, are localized on the host cell surface and at the endosomal compartment or both cell surface and endosomes, depending on the cell type. For recognition and activation of TLR4 by LPS, a set of adaptor proteins, MD2 and CD14, are necessary. These adaptor proteins are located extracellularly. The MD2–TLR4 complex is able to distinguish smooth or rough forms of LPS [3], where CD14 relays the signal accordingly [4]. For the rough form, the signaling through TLR4 is MyD88-dependent, and when smooth LPS serves as ligand, the TRIF-mediated signaling cascade is the dominant form of downstream gene activation [4]. The distinction between smooth and rough form is based on the oligosaccharide component of the LPS [5, 6]. LPS on the surface of *Salmonella* is of the smooth form, which suggests that the TRIF pathway is the predominant mechanism of type I IFN expression. The lipid A component of LPS is also highly inflammatory, which activates the MyD88 pathway, and synthetic structures with modification of lipid A have been shown to selectively induce the TRIF pathway [7]. Furthermore, the modifications of lipid A also determine how potent if any the signaling cascades are activated. Namely, LPS comprised of hexa- and hepta-acetylated lipid A is strongly

B. Shutinoski • S. Sad (✉)
Faculty of Medicine, Department of Biochemistry, Microbiology and Immunology, University of Ottawa, 451 Smyth Road, Ottawa, ON, Canada K1H 8M5
e-mail: subash.sad@uottawa.ca

© Springer International Publishing Switzerland 2014

D. Parker (ed.), *Bacterial Activation of Type I Interferons*,
DOI 10.1007/978-3-319-09498-4_7

inflammatory, in contrast to tetra-acetylated lipid A [8, 9]. The pathogenic *Salmonella* LPS has the smooth oligosaccharide component and the hexa-acetylated lipid A. This assures that the host cells are strongly engaged and potent inflammatory response is mounted. Type I interferon production in response to TLR4 engagement occurs predominantly through a MyD88 independent, TRIF-dependent mechanism [10].

Recent studies have shown that *Salmonella* exploits this induction of a strong inflammatory response to promote its intracellular survival [11, 12]. CpG treatment of mice that normally resolve *S. typhimurium* infection resulted in host susceptibility [12]. This was due to the enhanced intracellular proliferation of *Salmonella*, which requires expression of the *Salmonella* pathogenicity island 2 (SPI-2) genes [12]. In another study, when TLR2-TLR4-TLR9 triple knock-out mice were infected with *Salmonella*, they survived better than combinations of double knock-outs of the same TLR members [11]. Again this was shown to operate through induction of SPI-2 genes, which were induced in response to TLR engagement. Activated TLR9 recruits MyD88, IRAK1, IRAK4, and TRAF6 to phosphorylate/activate IRF7, followed by IRF7 translocation in the nucleus where it can activate type I IFNs production [13]. This is summarized in Fig. 1.

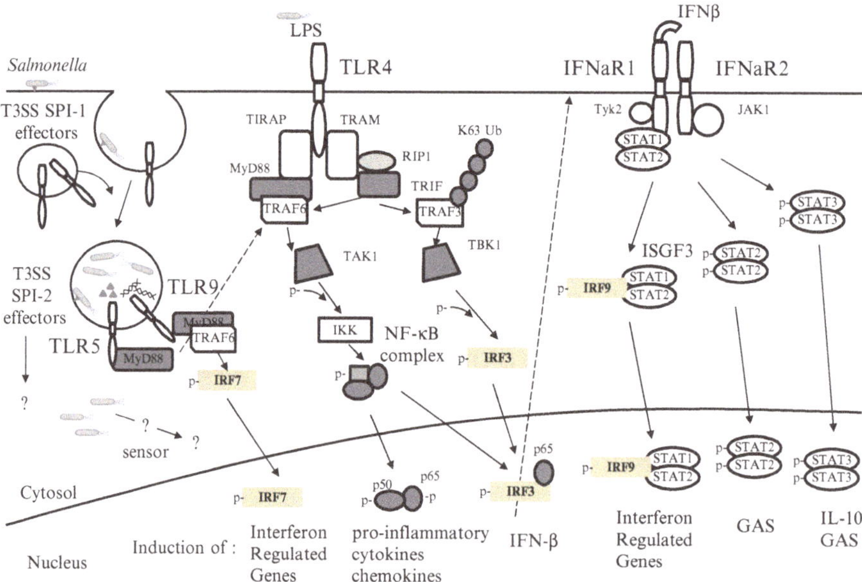

Fig. 1 Major induction pathways of type I IFNs by *Salmonella* and role of type I IFN during infection at the cellular level. *Salmonella* SPI-1 effectors induce its engulfment in a SCV or phagosome where the SPI-2 effectors get induced. Although many functions are described for SPI-2 effectors, it remains unclear whether they regulate type I IFN production. Once in the phagosome or SCV, activated TLR9 can relay signals to IRF7 to stimulate IRGs and the TLR5 similar to TLR4 via MyD88 pathway activates the NF-κB complex. LPS activated TLR4, signals through MyD88 or TRIF-dependent pathways. The MyD88 pathway, via activated NF-κB, leads to induction of proinflammatory cytokines and chemokines and the TRIF pathway leads to IRF3 activation and type I IFN production. The type I IFN produced then engages the IFNAR to induce production of several hundreds of interferon regulated genes or gamma-activated sequences (GAS) via autocrine loop. *IKK* inhibitor of NF kappa-B kinase, *TRAF* tumor necrosis factor receptor-associated factor

Type III Secretion System-Dependent Induction of Type I IFN

Salmonella infects various types of cells. While phagocytic cells such as macrophages and dendritic cells can rapidly phagocytose *Salmonella*, the non-phagocytic cells are infected through a type III secretion system (T3SS) encoded in the SPI-1 cluster of genes. The T3SSs are needle-like structures canonically used by bacteria to bridge bacterial cytoplasm with the host cytosol and translocate proteinaceous effector molecules, which in case of pathogenic bacteria subvert host cell signaling [14]. The SPI-1 induces host cell structures that promote engulfment of *Salmonella* and its intracellular translocation into vacuoles, termed *Salmonella* containing vacuoles, SCVs. Professional phagocytic cells don't require SPI-1 to phagocytose *Salmonella*, and once intracellular, the host could potentially recognize other pathogen-associated molecular patterns (PAMPs) beside LPS. However, SPI-1 is active in the phagocytic cells as well. PrgJ, a capping protein of the T3SS of SPI-1, gets removed from the needle structure of T3SS and enables secretion of *Salmonella* effectors. This allows *Salmonella* to engage the NLRC4 inflammasome [15]. Flagellin, which is expressed by *Salmonella*, and is needed for its virulence, serves as a signal for TLR5 and NLRC4 inflammasome engagement, which in turn leads to activation and production of proinflammatory cytokines [16]. The flagellum is evolutionary related to the T3SS machinery and in certain conditions can secrete proteins as well [17, 18]. By engaging the inflammasomes, the production of active IL-1β is maintained, which is able to positively feed into the type I IFN production by inhibiting the DUBA, deubiqutinase known to remove K63 ubiquitination of TRAF3 [19]. K63 ubiquitination of TRAF3 is a major modification required for IFN gene expression [20].

In a study that addressed the role of caspase-8 during *Salmonella* infection it was shown that caspase-8 is recruited to the inflammasome complex. This recruitment was shown to be specific to *S. typhimurium* infection and as part of that complex it contributed positively to IL-1β production [21]. The production of active IL-1β seems to be fine-tuned, as it is shown that SipB, a *Salmonella* SPI-1 effector protein, promotes its production [22]. Active IL-1β has many other functions, yet the IL-1R signaling by modulating TRAFs remains instrumental for type I IFN production [13, 20]. It is important to note that IL-1 signaling can also accelerate the degradation of IFNAR by activating kinases that add phospho-moiety to a so-called degron sequence within the IFNAR protein [23], therefore adding complexity to the role of IL-1 signaling in type I IFN production and signaling.

Microarray studies focused on the host response to *Salmonella* infection revealed that many genes are specifically activated. RAW24.7, a murine macrophage cell line infected by *S. typhimurium*, was assessed for gene expression. The following genes were found to be upregulated: MIP-1α, MIP-1β, MIP-2α, IL-1β, TNF receptor, CD40, IκBα, IκBβ, NF-E2, IRF1, and c-rel among many [24]. In a similar study it was shown that SPI-1 effectors exploit host pathways that are independent of TLR engagement. Many genes in uninfected control remained at same expression level as cells infected by SPI-1 mutant *Salmonella* strain [25]. In that same study STAT3, a transcriptional factor with pleiotropic effects was upregulated [25], which in cooperation with IRF1 regulated the production of IL-10 [26]. Indeed, IL-10 is an anti-inflammatory cytokine, which has been shown to promote the intracellular proliferation of *Salmonella* [27].

Role of Type I IFNs During *Salmonella* Infection

In various infectious disease models (e.g., *Listeria*, *Mycobacteria*, *Trypanosoma*, *Candida*), it has been shown that IFNAR-deficient mice display enhanced survival [28–31]. Similarly, IFNAR-deficient mice display enhanced survival during infection with virulent *S. typhimurium* [32]. It is conceivable that pathogens from different domains of life or classes have converged in utilizing mechanisms of subverting the host immune defenses, and the above-mentioned examples would reiterate the importance of type I IFN signaling in host–pathogen interactions. The complexity of interferon signaling pathways and its impact on *Salmonella* pathogenesis was further revealed in another study in which UBP43-deficient mice (alternatively known as USP18) were shown to have elevated type I interferon signaling, yet these mice were able to control *Salmonella* better in vivo, since the splenic bacterial burden was reduced in UBP43-deficient mice; however, there was no difference in host susceptibility between WT and UBP43-deficient mice [33, 34]. UBP43 is a member of the "Ubiquitin specific protease" family that cleaves ISG15, a ubiquitin-like posttranslational modification (PTM) of proteins, which appears to be dependent on IFN-signaling [35]. The mechanism behind the better control of *Salmonella* in UBP43-deficient mice was attributed to the sustained and hyperactive JAK-STAT1 signaling, as the failure to remove ISG15 from the JAK1 resulted in prolonged JAK1-STAT1 signaling [36]. Furthermore, UBP43-deficient mice displayed elevated expression of genes that are dependent on type I IFN signaling (ISGs), and were hypersensitive to LPS-induced septic shock [33]. While these results may appear to be at odds with the phenotype obtained in IFNAR-deficient mice, however, the UBP43 deficient mice display elevated inflammatory signaling in contrast to IFNAR-deficient mice. Elevated inflammatory signaling in UBP43-deficient mice may promote initial clearance of bacteria, but the overt inflammatory response may lead to fatality at a later time period. Work on *Salmonella* invasiveness after treatment with type I IFN, suggests that epithelial cells are less susceptible to invasion [37], and because of that impaired invasion it is argued that mice challenged intragastrically with *Salmonella* show enhanced survival if treated with type I IFNs [38].

Furthermore type I IFN signaling is implicated in the regulation of inflammasome activation, and stimulation of necrosome formation, both presently understood as distinct signaling complexes. Inflammasomes are protein complexes that enable activation of inflammatory caspases, which drive immune responses by stimulating the production of proinflammatory cytokines, and by inducing pyroptosis, a mechanism of proinflammatory cell death [39]. Work done on elucidating the mechanisms involved in inflammasome regulation by IFNAR signaling indicated that type I IFN inhibits the production of IL-1β, through regulation of the NLRP3, leading to reduced transcript levels of pro-IL-1β [40]. Yet still, during infection with gram-negative bacteria, type I IFN promotes IL-1β production by controlling caspase-11 activity [41], and most likely such duality is dependent on the amount of IFN-β.

Necrosome is a protein complex that when assembled leads the host cell to necroptosis, a proinflammatory mechanism of cell death. Typically it is induced by TNFα-TNFR1 interaction in the absence of apoptosis [42]. During *S. typhimurium* infection of macrophages it was shown that type I IFN signaling stimulates necrosome activation leading to necrotic cell death, where IFNAR KO bone marrow macrophages showed enhanced survival [32]. Type I IFN signaling is the critical check-point of necrosome activation in macrophages. During in vivo infection, IFNAR-deficient mice had more macrophages, which correlated to better control of *Salmonella*. Additionally, the abrogated cytokine signaling downstream of IFNAR can also be a contributing factor, as the pleiotropic effects of IFN signaling can modulate subsequent downstream cytokine and chemokine signaling. Necroptosis is induced by IFN-α/β and IFN-γ signaling pathways independent of death receptors signaling, but dependent on Protein kinase RNA-activated (PKR) and Fas-associated death domain (FADD) [43]. Further, even TNF-dependent necrosome activation appears to be dependent on type I IFN signaling (S. Sad, unpublished).

A hallmark of necroptosis is the release of damage associated molecular patterns (DAMPs) that can act as "secondary" ligands during host–pathogen interactions and can become major drivers of inflammatory responses, although their contribution is often neglected. Necrosome activation that is associated with *Salmonella* infection that is notorious for inducing host cell death generates overt pathology leading to adverse outcome. During infections by pathogens that are able to inhibit caspases, necroptosis can be regarded as a backup mechanism that initiates inflammatory cell death and alerts the immune system defenses. Specifically, in *Salmonella* infection, the outcome and progression are multifactorial and will not be only dependent on type I IFN signaling [44], yet the IFNAR-deficient mice show significantly reduced susceptibility to *Salmonella* infection [32].

Final Remarks

New pathways of type I IFN signaling have emerged that seem to indicate that the impact of type I IFN signaling may be highly dependent on the disease context [45]. The IFNAR KO mice have been used extensively in many studies and have revealed both the positive and negative role of type I IFN signaling. At the cellular level the role of type I IFN signaling is also complex. Resistance to LPS shock is mediated by ablation of type I IFN signaling, as IFNAR1 KO, but not the IFNAR2 KO, mice are resistant to LPS [45, 46]. Type I IFN appears to be a key mechanism that impacts inflammasome and necrosome activation, although the precise mechanistic details are lacking currently. These two distinct signaling complexes, inflammasome and necrosome, might have substantial cross-talk since both are controlled by type I IFN. *Salmonella* is a chronic intracellular pathogen, which results in persistent activation of immune response. It is therefore quite conceivable that type I IFN signaling plays a key role in this process, which results in a deleterious host outcome due to persistent pathology.

References

1. Broz P, Monack DM (2013) Newly described pattern recognition receptors team up against intracellular pathogens. Nat Rev Immunol 13:551–565
2. Iwasaki A, Medzhitov R (2010) Regulation of adaptive immunity by the innate immune system. Science 327:291–295
3. Huber M, Kalis C, Keck S, Jiang Z, Georgel P, Du X, Shamel L, Sovath S, Mudd S, Beutler B, Galanos C, Freudenberg MA (2006) R-form LPS, the master key to the activation of TLR4/MD-2-positive cells. Eur J Immunol 36:701–711
4. Jiang Z, Georgel P, Du X, Shamel L, Sovath S, Mudd S, Huber M, Kalis C, Keck S, Galanos C, Freudenberg M, Beutler B (2005) CD14 is required for MyD88-independent LPS signaling. Nat Immunol 6:565–570
5. Nikaido H (2003) Molecular basis of bacterial outer membrane permeability revisited. Microbiol Mol Biol Rev 67:593–656
6. Wilkinson SG (1996) Bacterial lipopolysaccharides–themes and variations. Prog Lipid Res 35: 283–343
7. Bowen WS, Minns LA, Johnson DA, Mitchell TC, Hutton MM, Evans JT (2012) Selective TRIF-dependent signaling by a synthetic toll-like receptor 4 agonist. Sci Signal 5:ra13
8. Janusch H, Brecker L, Lindner B, Alexander C, Gronow S, Heine H, Ulmer AJ, Rietschel ET, Zahringer U (2002) Structural and biological characterization of highly purified hepta-acyl lipid A present in the lipopolysaccharide of the *Salmonella enterica* sv. Minnesota Re deep rough mutant strain R595. J Endotoxin Res 8:343–356
9. Lapaque N, Takeuchi O, Corrales F, Akira S, Moriyon I, Howard JC, Gorvel JP (2006) Differential inductions of TNF-alpha and IGTP, IIGP by structurally diverse classic and non-classic lipopolysaccharides. Cell Microbiol 8:401–413
10. Toshchakov V, Jones BW, Perera PY, Thomas K, Cody MJ, Zhang S, Williams BR, Major J, Hamilton TA, Fenton MJ, Vogel SN (2002) TLR4, but not TLR2, mediates IFN-beta-induced STAT1alpha/beta-dependent gene expression in macrophages. Nat Immunol 3:392–398
11. Arpaia N, Godec J, Lau L, Sivick KE, McLaughlin LM, Jones MB, Dracheva T, Peterson SN, Monack DM, Barton GM (2011) TLR signaling is required for *Salmonella typhimurium* virulence. Cell 144:675–688
12. Wong CE, Sad S, Coombes BK (2009) *Salmonella enterica* serovar typhimurium exploits Toll-like receptor signaling during the host-pathogen interaction. Infect Immun 77:4750–4760
13. Gonzalez-Navajas JM, Lee J, David M, Raz E (2012) Immunomodulatory functions of type I interferons. Nat Rev Immunol 12:125–135
14. Cornelis GR (2006) The type III secretion injectisome. Nat Rev Microbiol 4:811–825
15. Miao EA, Mao DP, Yudkovsky N, Bonneau R, Lorang CG, Warren SE, Leaf IA, Aderem A (2010) Innate immune detection of the type III secretion apparatus through the NLRC4 inflammasome. Proc Natl Acad Sci U S A 107:3076–3080
16. Lage SL, Buzzo CL, Amaral EP, Matteucci KC, Massis LM, Icimoto MY, Carmona AK, D'Imperio Lima MR, Rodrigues MM, Ferreira LC, Amarante-Mendes GP, Bortoluci KR (2013) Cytosolic flagellin-induced lysosomal pathway regulates inflammasome-dependent and -independent macrophage responses. Proc Natl Acad Sci U S A 110:E3321–E3330

17. Karlinsey JE, Tanaka S, Bettenworth V, Yamaguchi S, Boos W, Aizawa SI, Hughes KT (2000) Completion of the hook-basal body complex of the *Salmonella typhimurium* flagellum is coupled to FlgM secretion and fliC transcription. Mol Microbiol 37:1220–1231
18. Paul K, Erhardt M, Hirano T, Blair DF, Hughes KT (2008) Energy source of flagellar type III secretion. Nature 451:489–492
19. Tseng PH, Matsuzawa A, Zhang W, Mino T, Vignali DA, Karin M (2010) Different modes of ubiquitination of the adaptor TRAF3 selectively activate the expression of type I interferons and proinflammatory cytokines. Nat Immunol 11:70–75
20. Hacker H, Tseng PH, Karin M (2011) Expanding TRAF function: TRAF3 as a tri-faced immune regulator. Nat Rev Immunol 11:457–468
21. Man SM, Tourlomousis P, Hopkins L, Monie TP, Fitzgerald KA, Bryant CE (2013) *Salmonella* infection induces recruitment of caspase-8 to the inflammasome to modulate IL-1beta production. J Immunol 191:5239–5246
22. Hersh D, Monack DM, Smith MR, Ghori N, Falkow S, Zychlinsky A (1999) The *Salmonella* invasin SipB induces macrophage apoptosis by binding to caspase-1. Proc Natl Acad Sci U S A 96:2396–2401
23. Fuchs SY (2013) Hope and fear for interferon: the receptor-centric outlook on the future of interferon therapy. J Interferon Cytokine Res 33:211–225
24. Rosenberger CM, Scott MG, Gold MR, Hancock RE, Finlay BB (2000) *Salmonella typhimurium* infection and lipopolysaccharide stimulation induce similar changes in macrophage gene expression. J Immunol 164:5894–5904
25. Bruno VM, Hannemann S, Lara-Tejero M, Flavell RA, Kleinstein SH, Galan JE (2009) *Salmonella typhimurium* type III secretion effectors stimulate innate immune responses in cultured epithelial cells. PLoS Pathog 5:e1000538
26. Saraiva M, O'Garra A (2010) The regulation of IL-10 production by immune cells. Nat Rev Immunol 10:170–181
27. Nguyen T, Robinson N, Allison SE, Coombes BK, Sad S, Krishnan L (2013) IL-10 produced by trophoblast cells inhibits phagosome maturation leading to profound intracellular proliferation of *Salmonella enterica* typhimurium. Placenta 34:765–774
28. Chessler AD, Caradonna KL, Da'dara A, Burleigh BA (2011) Type I interferons increase host susceptibility to *Trypanosoma cruzi* infection. Infect Immun 79:2112–2119
29. Majer O, Bourgeois C, Zwolanek F, Lassnig C, Kerjaschki D, Mack M, Muller M, Kuchler K (2012) Type I interferons promote fatal immunopathology by regulating inflammatory monocytes and neutrophils during *Candida* infections. PLoS Pathog 8:e1002811
30. Manca C, Tsenova L, Freeman S, Barczak AK, Tovey M, Murray PJ, Barry C, Kaplan G (2005) Hypervirulent M. tuberculosis W/Beijing strains upregulate type I IFNs and increase expression of negative regulators of the Jak-Stat pathway. J Interferon Cytokine Res 25: 694–701
31. O'Connell RM, Saha SK, Vaidya SA, Bruhn KW, Miranda GA, Zarnegar B, Perry AK, Nguyen BO, Lane TF, Taniguchi T, Miller JF, Cheng G (2004) Type I interferon production enhances susceptibility to *Listeria monocytogenes* infection. J Exp Med 200:437–445
32. Robinson N, McComb S, Mulligan R, Dudani R, Krishnan L, Sad S (2012) Type I interferon induces necroptosis in macrophages during infection with *Salmonella enterica* serovar typhimurium. Nat Immunol 13:954–962
33. Kim KI, Malakhova OA, Hoebe K, Yan M, Beutler B, Zhang DE (2005) Enhanced antibacterial potential in UBP43-deficient mice against *Salmonella typhimurium* infection by up-regulating type I IFN signaling. J Immunol 175:847–854
34. Richer E, Yuki KE, Dauphinee SM, Lariviere L, Paquet M, Malo D (2011) Impact of Usp18 and IFN signaling in *Salmonella*-induced typhlitis. Genes Immun 12:531–543
35. Sgorbissa A, Brancolini C (2012) IFNs, ISGylation and cancer: Cui prodest? Cytokine Growth Factor Rev 23:307–314
36. Malakhova OA, Yan M, Malakhov MP, Yuan Y, Ritchie KJ, Kim KI, Peterson LF, Shuai K, Zhang DE (2003) Protein ISGylation modulates the JAK-STAT signaling pathway. Genes Dev 17:455–460

37. Bukholm G, Degre M (1983) Effect of human leukocyte interferon on invasiveness of *Salmonella* species in HEp-2 cell cultures. Infect Immun 42:1198–1202
38. Bukholm G, Berdal BP, Haug C, Degre M (1984) Mouse fibroblast interferon modifies *Salmonella typhimurium* infection in infant mice. Infect Immun 45:62–66
39. Lamkanfi M, Dixit VM (2012) Inflammasomes and their roles in health and disease. Annu Rev Cell Dev Biol 28:137–161
40. Guarda G, Braun M, Staehli F, Tardivel A, Mattmann C, Forster I, Farlik M, Decker T, Du Pasquier RA, Romero P, Tschopp J (2011) Type I interferon inhibits interleukin-1 production and inflammasome activation. Immunity 34:213–223
41. Rathinam VA, Vanaja SK, Waggoner L, Sokolovska A, Becker C, Stuart LM, Leong JM, Fitzgerald KA (2012) TRIF licenses caspase-11-dependent NLRP3 inflammasome activation by gram-negative bacteria. Cell 150:606–619
42. Vandenabeele P, Galluzzi L, Vanden Berghe T, Kroemer G (2010) Molecular mechanisms of necroptosis: an ordered cellular explosion. Nat Rev Mol Cell Biol 11:700–714
43. Thapa RJ, Nogusa S, Chen P, Maki JL, Lerro A, Andrake M, Rall GF, Degterev A, Balachandran S (2013) Interferon-induced RIP1/RIP3-mediated necrosis requires PKR and is licensed by FADD and caspases. Proc Natl Acad Sci U S A 110:E3109–E3118
44. Ramos-Morales F (2012) Impact of *Salmonella enterica* type III secretion system effectors on the eukaryotic host cell. ISRN Cell Biol 2012:32
45. de Weerd NA, Vivian JP, Nguyen TK, Mangan NE, Gould JA, Braniff SJ, Zaker-Tabrizi L, Fung KY, Forster SC, Beddoe T, Reid HH, Rossjohn J, Hertzog PJ (2013) Structural basis of a unique interferon-beta signaling axis mediated via the receptor IFNAR. Nat Immunol 14: 901–907
46. Mahieu T, Park JM, Revets H, Pasche B, Lengeling A, Staelens J, Wullaert A, Vanlaere I, Hochepied T, van Roy F, Karin M, Libert C (2006) The wild-derived inbred mouse strain SPRET/Ei is resistant to LPS and defective in IFN-beta production. Proc Natl Acad Sci U S A 103:2292–2297

Yersinia Activation of Type I Interferon

Miqdad O. Dhariwala and Deborah M. Anderson

Introduction

Three *Yersinia* species are pathogenic to mammals and are part of the Enterobacteriaceae family of eubacteria. *Yersinia enterocolitica* causes gastroenteritis, a normally self-limiting infection that has been associated with significant outbreaks of yersiniosis in humans and animals throughout the world [1, 2]. *Yersinia pseudotuberculosis* causes less severe infection and is less commonly associated with foodborne outbreaks but is more closely related to the deadly *Yersinia pestis*, which diverged approximately 3,000 years ago [3]. *Y. pestis* is the causative agent of bubonic plague, a flea borne disease which is characterized by a late stage bacteremia that seeds multiple tissues including the lungs [4]. *Y. pestis* infection of the respiratory tract leads to a fulminant bronchopneumonia which can be spread through respiratory secretions.

Enhanced virulence through systemic infection and the flea life cycle of *Y. pestis* are due to the acquisition of two plasmids as well as genetic reduction, loss of function mutations that in some cases reduced the activity of virulence factors conserved in the other *Yersinia* species [5]. Importantly, all three pathogens employ temperature-dependent changes in LPS composition such that the immunostimulatory hexacylated lipid A is down-regulated at the mammalian host temperature of 37 °C [6]. Tetraacylated LPS from *Yersinia* grown at 37 °C provides little to no stimulation of TNFα secretion. Hypoacetylation of *Y. pestis* LPS at 37 °C makes a significant contribution to virulence by providing evasion from toll-like receptor 4 (TLR4) signaling, thereby limiting the activation of NF-κB [7]. This structure also contributes to evasion of inflammasome activation [8]. Nevertheless, in spite of this and other immune evasive strategies used to control inflammatory responses,

M.O. Dhariwala • D.M. Anderson (✉)
Department of Veterinary Pathobiology, University of Missouri, Columbia, MO 65211, USA
e-mail: andersondeb@missouri.edu

© Springer International Publishing Switzerland 2014

D. Parker (ed.), *Bacterial Activation of Type I Interferons*,
DOI 10.1007/978-3-319-09498-4_8

Y. pestis still induces the production of type I interferon (IFN) [9]. In this chapter, we will review signaling pathways that may induce type I IFN and the phenotypic outcome of IFN signaling during *Yersinia* infection.

Extracellular Bacteria Evade Activation of Interferon Responses

Extracellular *Yersinia* target phagocytic cells for the injection of anti-host proteins, known as Yops, by the type III secretion system (T3SS) [10]. Yop activities result in inhibition of phagocytosis and production of proinflammatory cytokines and lead to host cell death in vitro [11] (Fig. 1). *Yersinia* that makes intimate contact with a host cell will insert a translocation pore into its plasma membrane through which effector Yops are transported. Pore formation can be detected by the cell's inflammasome machinery, which includes cytoplasmic nucleic acid pattern recognition receptors. In the absence of Yop effectors, MyD88- and TRIF-independent production of type I

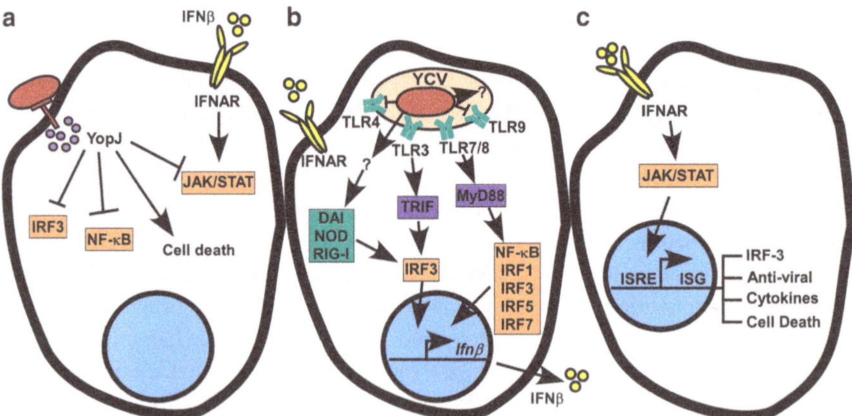

Fig. 1 *Yersinia* activation of type I IFN in macrophages and other phagocytic cells, with possible downstream signaling outcomes. (**a**) Macrophages making intimate contact with *Yersinia* (*red*) are injected with Yops by the T3SS. One effector, YopJ blocks activation of major pro-inflammatory signaling pathways mediated by IRF3, NF-κB, and p38 kinase, and promotes apoptosis. Signaling from IFNAR (*yellow*) is probably blocked in these cells, but this has not yet been directly shown. (**b**) Intracellular bacteria reside in a *Yersinia* containing vacuole (YCV) where they may prevent signaling by TLR4 and TLR9. In addition, YCV may secrete proteins or other products into the cytoplasm or vacuole lumen that might activate one or more intracellular PRRs (*blue*, DAI, NOD, RIG-I or TLR3, TLR7, respectively). Signaling can be through MyD88, TRIF (*purple*) or independent of either, leading to induction of one or more transcription factors (*orange*) and the expression of IFN-β. Cells carrying intracellular *Yersinia* may also respond to IFN-β through IFNAR in an unknown manner. (**c**) Uninfected immune cells residing in different sites such as the bone marrow, may also respond to IFN-β through IFNAR, inducing interferon stimulated gene (ISG) expression that may upregulate an anti-viral response, IRF3, inflammatory cytokines, and/or cell death genes

IFN were observed following insertion of the type III translocation pore by *Y. pseudotuberculosis* [12]. MyD88-/TRIF-independent induction of IFN was not blocked by the inhibition of phagocytosis suggesting that extracellular, rather than intracellular, bacteria were detected by macrophages. Deletion of the type III secretion system translocator YopB resulted in no detectable type I IFN production in vitro. These data suggest that host sensing of insertion of the type III translocation pore leads to expression of IFN-β through a cytoplasmic receptor. One or more Yop effectors may prevent expression of IFN-β either within the cell or by producing anti-inflammatory molecules that prevent cells from responding to cytokine signals.

YopJ/P is a deubiquitinase and an acetylase that primarily targets MAP2 kinases, and prevents activation of master inflammatory regulators such as NF-κB, AP-1, and IRF3 (Fig. 1a) [13–16]. During infection of macrophages in vitro, YopJ/P inactivation of NF-κB not only prevents production of proinflammatory cytokines, but it also induces apoptosis due to suppression of the NF-κB-dependent expression of anti-apoptotic proteins Bid and t-Bid [17]. Given these activities, it seems likely that YopJ injection would prevent the expression of IFN-β either due to acetylation of IRF3, NF-κB or even another cytoplasmic target required for signaling from the translocation pore. Even though deletion of *yopJ* results in loss of immune suppression of macrophages in vitro, there appears to be only a small contribution of YopJ to virulence in mouse models of disease [18, 19].

Yersinia Evades TLR4 Activation

TLR4 is a host pattern recognition receptor that in conjunction with MD-2 can recognize the Gram-negative bacterial cell membrane component LPS [20–22]. Activation of TLR4 can occur at the plasma membrane or in the endosome, with different outcomes [23]. Upon binding LPS at the plasma membrane, the scaffold protein TIRAP (MAL) is recruited via its TIR domain. TIRAP binding results in recruitment of MyD88, which leads to the activation of the transcription factor NF-κB and the production of proinflammatory cytokines such as TNFα and IL-6. Alternatively, when TLR4 activation occurs in the endosomal compartment after, for example, the phagocytosis of Gram-negative bacteria, the downstream adapter molecule TRIF mediates the activation of the transcription factor IRF3 and induces the expression of IFN-β.

Yet, it appears that TLR4 signaling does not play a major role during *Y. pestis* infection, as $Tlr4^{-/-}$ mice challenged with wild-type *Y. pestis* are not more susceptible to infection [7]. All three *Yersinia* species modulate their lipid A structure in response to temperature which is believed to result in poor stimulation of TLR4 during infection. The TLR4–MD-2 complex binds hexaacylated lipid A but does not recognize hypoacetylated forms [23]. Hexaacylated lipid A is the most common form of Gram-negative LPS and the predominant species when *Yersiniae* are growing in the flea or environment. At 37 °C, the bacterial genes encoding lipid IV acetylases LpxL and LpxM are down-regulated and newly synthesized LPS incorporates tetraacylated lipid A. While the hexacylated lipid A that predominates at lower temperatures stimulates TLR4, the tetraacylated form does not stimulate TLR4 on human or murine macrophages [24]. All three *Yersinia* species down-regulate the expression of hexaacylated lipid A at 37 °C and even though each expresses a unique LPS structure, none of these is stimulatory to TLR4 [25]. In summary, *Y. enterocolitica*, *pseudotuberculosis*, and *pestis* share dominant virulence factors of hypoacetylated LPS and injection of immunomodulatory Yops by the T3SS. These virulence factors combine to provide for evasion of TLR4 and the expression of pro-inflammatory cytokines during infection. Yet in spite of these dominant immune modulatory mechanisms, host type I IFN signaling pathways appear to be active during infection.

Susceptibility of Mice to Yersiniosis and Plague is Affected by Type I Interferon Signaling

TRIF is required during *Y. enterocolitica* infection and its signaling led to protective responses in a murine model of yersiniosis. *Trif*$^{-/-}$ mice were found to be defective in phagocytosis of bacteria, which may have contributed to an increase in bacterial dissemination and elevated titers [26]. *Trif*$^{-/-}$ mice produced reduced amounts of IFN-β and IFN-γ during *Y. enterocolitica* infection suggesting that both cytokines depended on TRIF signaling for production. Although IFN-γ activates macrophages to up-regulate bactericidal mechanisms, it is unclear if this would be sufficient to have an impact on the infection since the closely related *Y. pestis* resists killing by activated macrophages [27, 28]. In addition, previous work showed a requirement for TRIF in inducing apoptosis in macrophages following infection by *Y. enterocolitica* [29]. TLR4 was required for TRIF-dependent apoptosis but not MyD88. Together these data suggest that *Y. enterocolitica* may induce type I IFN expression through TLR4 activation of TRIF from the phagosomal membrane rather than the cell surface. This hypothesis is at odds with the observation that LPS from *Y. enterocolitica* does not induce inflammatory cytokine production, and additional work is needed to understand how TRIF is stimulated during infection. Whether the TRIF-dependent cell death, phagocytosis or inflammatory responses required type I IFN signaling has not yet been reported. Furthermore, the phenotype of *Trif*$^{-/-}$ mice in a plague infection model has not yet been reported and it is unclear if host defense against other *Yersinia* would also require TRIF.

IRF3 is a transcription factor that is activated downstream of the adaptors TRIF, MyD88, or STING leading to expression of *Ifnb* and other interferon-stimulated genes. Although the sensitivity of *Trif*$^{-/-}$ mice to plague has not yet been reported, *Irf3*$^{-/-}$ mice have been tested in a murine model of pneumonic plague. *Irf3*$^{-/-}$ mice were more susceptible to the non-pigmented mutant *Y. pestis*, but not fully virulent bacteria [9]. Similar to *Trif*$^{-/-}$ mice, *Irf3*$^{-/-}$ macrophages were defective for phagocytosis of *Y. pestis* which may have contributed to accelerated growth and progression of the infection. However, bone marrow derived macrophages from *Ifnar*$^{-/-}$ mice were not defective for phagocytosis indicating that type I IFN is not involved this process. In fact, *Ifnb* expression was found in infected lung homogenate in WT and *Irf3*$^{-/-}$ mice suggesting that IRF3 is not required for *Ifnb* expression. Furthermore, unlike the *Trif*$^{-/-}$ mice infected with *Y. enterocolitica*, *Ifng* expression was not dependent on IRF3 and was absent until late stage infection in the presence or absence of IRF3. These results do not rule out signaling through IRF3 or TRIF as contributing to the type I IFN response, and additional data is needed to identify the signaling cascade induced by *Yersinia* for expression of type I IFN.

In striking contrast to *Irf3*$^{-/-}$ mice, *Ifnar*$^{-/-}$ mice were more resistant to plague suggesting that IFN-β signaling is immunopathogenic [9]. No changes in pro-inflammatory cytokines were associated with *Ifnar*$^{-/-}$ mice. Instead, IFN-dependent sensitivity to infection manifested under conditions of high bacterial burden, where the IFNAR-expressing mice lost control over bacterial growth while the *Ifnar*$^{-/-}$ mice cleared the infection. Neutrophil populations appeared depleted in the bone marrow and periphery of WT mice which may have led to their poor outcome. Together the data suggested that IFN signaling during *Y. pestis* infection caused an increase in neutrophil cell death or a decrease in maturation of cells in the bone marrow. The sensitivity of *Ifnar*$^{-/-}$ mice to *Y. enterocolitica* has not yet been reported and it remains unclear if a similar mechanism of neutrophil depletion is a common feature of the type I IFN response to *Yersinia*.

Activation of Intracellular PRRs by Intracellular *Yersinia*?

Although *Yersinia* has a predominantly extracellular life cycle, it also has the ability to invade phagocytic and non-phagocytic cells and will survive and grow in macrophages, eventually causing their death [30, 31, 27, 32]. The T3SS is not required for the intracellular life cycle, being only weakly expressed, and the role of hypoacylated LPS in the intracellular compartment has not been rigorously examined. Following phagocytosis or invasion, bacteria have been observed to localize in spacious vacuoles known as *Yersinia* containing vacuoles (YCV) which have cell surface markers found on late endosomes and autophagosomes (Fig. 1b) [33]. Survival in the YCV depends on the bacterial 2-component signaling pathway PhoPQ which is activated in low magnesium or low pH environments [34]. PhoP is also required for virulence of *Y. pestis* suggesting that intracellular survival is important to pathogenesis [35]. Little evidence has been presented supporting cytoplasmic localization of *Yersinia*. However, genome annotation of all three *Yersinia* pathogens has revealed the presence of multiple secretion systems, some of which could be utilized in the intracellular compartment to facilitate nutrient uptake or escape from the vacuole or cell. Overall, these data suggest that intracellular bacteria could be detected by cytoplasmic or endosomal PRRs, any number of which could result in expression of type I IFN.

While TLR4 is a major PRR for detection of bacteria that leads to type I IFN expression, TLR3 is activated primarily by viruses and double-stranded RNA in phagolysosomes, leading to type I IFN expression and an anti-viral response. Evidence suggests that TLR3 may also recognize bacteria. Gut epithelial cells expressing TLR3 can be stimulated by Gram-positive bacteria in the microbiota resulting in an anti-viral response that may serve to suppress the inflammatory response to commensal bacteria [36, 37]. TLR3 activation as a host defense mechanism against bacteria has not yet been reported but it is nevertheless clear that bacteria can be recognized by TLR3. *Y. pestis*-derived tetraacylated LPS has previously been shown to reduce activation of signaling through the TLR2 and TLR3 pathways [38]. This data demonstrates the ability of *Y. pestis* to actively suppress the innate immune response through receptor crosstalk and suggests that TLR3 may also be neutralized during infection due to the *Yersinia* LPS structure.

Other toll-like receptors, such as TLR7 and TLR9, also localize to the phagosomal membrane where they can be activated upon recognition of microbial nucleic acids. TLR7 and TLR9 signal downstream to MyD88 to activate NF-κB, IRF1, IRF3, IRF5, and/or IRF7 to induce the expression of IFN-β [39–43]. TLR9 binds unmethylated dinucleotides to induce a downstream signal transduction pathway involving MyD88 and resulting in activation of *Ifnb* expression [23]. Active stimulation of TLR9 prior to infection improved clearance of non-pigmented *Y. pestis* in a murine respiratory infection model [44]. Since IFN-β signaling was previously associated with immunopathology, these data suggest that the protective effect of TLR9 signaling may not be related to IFN-β and TLR9 may not be principally responsible for *Ifnb* expression. Furthermore, these data suggest that *Y. pestis* may even prevent TLR9 activation. All three *Yersinia* pathogens have similar capability for intracellular survival and replication and there is evidence that live bacteria prevent acidification of the phagolysosome, which is necessary for the localization and activation of TLR7 in this compartment [33]. Together, the data suggest that live *Yersiniae* are likely to prevent signaling from the nucleic acid sensors of the phagosome. It may be that bacteria that lyse in a small percentage of macrophages are detected by nucleic acid PRRs and the resulting type I IFN signal is amplified by neighboring cells (Fig. 1c). Furthermore, intracellular *Y. pestis* eventually kill their host cells in an active process that requires intracellular survival. Perhaps the mechanism that is used to cause host cell death from the phagosome also induces expression of IFN-β from cytoplasmic PRRs. Overall, the mechanism underlying *Yersinia* activation of type I interferon remains incompletely understood and may be a critical part of its pathogenesis.

Concluding Remarks

Yersinia are potentially recognized throughout their infectious life cycle from an early intracellular phase to later anti-phagocytic, rapid growth phase and therefore are likely to interact with PRRs on the plasma and phagosomal membranes as well as within the cytoplasm [32]. Although all three species of pathogenic *Yersinia* have in common two major virulence factors, an atypical LPS that is not stimulatory to TLR4 and the T3SS, only one causes severe sepsis with multi-organ failure while the others cause self-limiting gastroenteritis. The differences are presumably caused at least in part by different host responses to infection, originating with PRRs or downstream adaptor molecules. All three *Yersinia* species that are pathogenic to mammals suppress PAMPs at 37 °C that would otherwise stimulate activation of TLR2, -3, -4, -5, and possibly -9, thereby disabling recognition of intracellular and extracellular bacteria [45].

Nevertheless, *Yersinia* are detected by the mammalian innate immune system and mice that lack components of type I IFN signaling pathways have altered sensitivity to infection. The complex *Yersinia* lifecycle in the mammalian host includes the display of many PAMPs: insertion of the type III translocation pore, injection of bacterial proteins in the host cytoplasm, modification of intracellular trafficking to permit growth and survival within the YCV, and promoting escape of intracellular bacteria by host cell lysis. PRRs at the plasma membrane, phagosomal membrane and even cytoplasm have the opportunity to see *Yersinia* during infection.

The adaptor TRIF is required for *Yersinia* YopJ-induced apoptosis through caspase 8 and 9 and it is tempting to speculate that TRIF-dependent type I IFN signaling contributes to the control of apoptosis or other forms of programmed cell death as it does following viral infection [29]. During *Salmonella* infection, type I IFN signaling led to activation of necroptosis in infected macrophages which enhanced virulence [46]. The bacterial and host proteins that were responsible for this were not identified leaving it unknown whether *Yersinia* infection could have a similar effect on neutrophils, macrophages or even hematopoetic precursor cells. For the plague model, IFN pathology manifests during late stage disease when bacteria have spread to distal sites where they grow logarithmically. Thus it may well be that only infected cells are responding poorly to type I IFN.

Type I IFN is used as a therapeutic to induce anti-viral and anti-cancer mechanisms in humans and the data gathered to date on the role of type I IFN during *Yersinia* infection suggests that this type of treatment could generate an increased risk of disease [47, 48]. Conversely, it seems likely that blocking type I IFN signaling might improve the outcome of late stage plague. Perhaps single ISGs are responsible for IFN-related pathology and could be specifically targeted as an anti-plague therapeutic. Given the dependence of viral clearance on the type I IFN response, it would be preferred to target one or a few ISGs as this would be less likely to generate an increase in susceptibility to viruses. Future experiments to identify ISGs associated with pathology or host defense against *Yersinia* may result in important advances in interferon therapies for humans.

References

1. Bottone E (1999) *Yersinia enterocolitica*: overview and epidemiologic correlates. Microbes Infect 1:323–333
2. McNally A, Cheasty T, Fearnley C, Dalziel R, Paiba G, Manning G, Newell D (2004) Comparison of the biotypes of *Yersinia enterocolitica* isolated from pigs, cattle and sheep at slaughter and from humans with yersiniosis in Great Britain during 1999–2000. Lett Appl Microbiol 39:103–108
3. Cui Y, Yu C, Yan Y, Lib D, Lia Y, Jombart T, Weinert L, Wang Z, Guo Z, Xu L, Zhang Y, Zheng H, Qin N, Xiao X, Wu M, Wang X, Zhou D, Qi Z, Du Z, Wu H, Yang H, Cao H, Wang H, Wang J, Yao S, Rakin A, Li Y, Falush D, Balloux F, Achtman M, Song Y, Wang J, Yang R (2013) Historical variations in mutation rate in an epidemic pathogen, *Yersinia pestis*. Proc Natl Acad Sci U S A 110(2):577–582
4. Pollitzer R (1954) Plague. World Health Organization, Geneva
5. Chouikha I, Hinnebusch BJ (2012) Yersinia–flea interactions and the evolution of the arthropod-borne transmission route of plague. Curr Opin Microbiol 15(3):239–246
6. Rebeil R, Ernst RK, Jarrett CO, Adams KN, Miller SI, Hinnebusch BJ (2006) Characterization of late acyltransferase genes of *Yersinia pestis* and their role in temperature-dependent lipid A variation. J Bacteriol 188(4):1381–1388
7. Montminy S, Khan N, McGrath S, Walkowicz M, Sharp F, Conlon J, Fukase K, Kusumoto S, Sweet C, Miyake K, Akira S, Cotter R, Goguen J, Lien E (2006) Virulence factors of *Yersinia pestis* are overcome by a strong lipopolysaccharide response. Nat Immunol 7(10):1066–1073
8. Valdimer G, Weng D, Paquette S, Vanaja S, Rathinam V, Aune M, Conlon J, Burbage J, Proulx M, Liu Q, Reed G, Mecsas J, Iwakura Y, Bertin J, Goguen J, Fitzgerald K, Lien E (2012) The NLRP12 inflammasome recognizes *Yersinia pestis*. Immunity 37(1):96–107
9. Patel A, Lee-Lewis H, Hughes-Hanks J, Lewis C, Anderson D (2012) Opposing roles for interferon regulatory factor-3 (IRF-3) and type I interferon signaling during plague. PLoS Pathog 8(7):e1002817
10. Marketon M, DePaolo R, DeBord K, Jabri B, Schneewind O (2005) Plague bacteria target immune cells during infection. Science 309:1739–1741
11. Raymond B, Young J, Pallett M, Endres R, Clements A, Frankel G (2013) Subversion of trafficking, apoptosis, and innate immunity by type III secretion system effectors. Trends Microbiol 21(8):430–441
12. Auerbuch V, Golenbock D, Isberg R (2009) Innate immune recognition of *Yersinia pseudotuberculosis* type III secretion. PLoS Pathog 5(12):1000686
13. Mukherjee S, Keitany G, Li Y, Wang Y, Ball H, Goldsmith E, Orth K (2006) *Yersinia* YopJ acetylates and inhibits kinase activation by blocking phosphorylation. Science 312: 1211–1214
14. Sweet C, Conlon J, Golenbock D, Goguen J, Silverman N (2007) YopJ targets TRAF proteins to inhibit TLR-mediated NF-κB, MAPK and IRF3 signal transduction. Cell Microbiol 9(11):2700–2715
15. Paquette N, Conlon J, Sweet C, Rus F, Wilson L, Pereira A, Rosadini C, Goutagny N, Weber A, Lane W, Shaffer S, Maniatis S, Fitzgerald K, Stuart L, Silverman N (2012) Serine/threonine acetylation of TGFβ-activated kinase (TAK1) by *Yersinia pestis* YopJ inhibits innate immune signaling. Proc Natl Acad Sci U S A 109(31):12710–12715
16. Mittal R, Peak-Chew S, McMahon H (2006) Acetylation of MEK2 and IκB kinase (IKK) activation loop residues by YopJ inhibits signaling. Proc Natl Acad Sci U S A 103(49): 18574–18579
17. Monack D, Mecsas J, Ghori N, Falkow S (1997) *Yersinia* signals macrophages to undergo apoptosis and YopJ is necessary for this cell death. Proc Natl Acad Sci U S A 94(19): 10385–10390
18. Monack D, Mecsas J, Bouley D, Falkow S (1998) *Yersinia*-induced apoptosis in vivo aids in the establishment of a systemic infection of mice. J Exp Med 188(11):2127–2137

19. Lemaitre N, Sebbane F, Long D, Hinnebusch B (2006) *Yersinia pestis* YopJ suppresses tumor necrosis factor alpha induction and contributes to apoptosis of immune cells in the lymph node but is not required for virulence in a rat model of bubonic plague. Infect Immun 74(9): 5126–5131
20. Poltorak A, He X, Smirnova I, Liu M, Van Huffel C, Du X, Birdwell D, Alejos E, Silva M, Galanos C, Freudenberg M, Ricciardi-Castagnoli P, Layton B, Beutler B (1998) Defective LPS signaling in C3H/HeJ and C57BL/10ScCr mice: mutations in *Tlr4* gene. Science 282:2085–2088
21. Chow JC, Young D, Golenbock D, Christ W, Gusovsky F (1999) Toll-like receptor 4 mediates lipopolysaccharide-induced signal transduction. J Biol Chem 274(16):10689–10692
22. Shimazu R, Akashi S, Ogata H, Nagai Y, Fukudome K, Miyake K, Kimoto M (1999) MD-2, a molecule that confers lipopolysaccharide responsiveness on Toll-like receptor 4. J Exp Med 199(11):1777–1782
23. Kawai T, Akira S (2010) The role of pattern-recognition receptors in innate immunity: update on Toll-like receptors. Nat Immunol 11(5):373–384
24. Hajjar A, Ernst R, Fortuno ES 3rd, Brasfield A, Yam C, Newlon L, Kollmann T, Miller S, Wilson C (2012) Humanized TLR4/MD-2 mice reveal LPS recognition differentially impacts susceptibility to *Yersinia pestis* and *Salmonella enterica*. PLoS Pathog 8(10):e1002963
25. Pérez-Gutiérrez C, Llobet E, Llompart C, Reinés M, Bengoechea J (2010) Role of lipid A acylation in *Yersinia enterocolitica* virulence. Infect Immun 78(6):2768–2781
26. Sotolongo J, Espana C, Echeverry A, Siefker D, Altman N, Zaias J, Santaolalla R, Ruiz J, Schesser K, Adkins B, Fukata M (2011) Host innate recognition of an intestinal bacterial pathogen induces TRIF-dependent protective immunity. J Exp Med 208(13):2705–2716
27. Pujol C, Grabenstein J, Perry R, Bliska J (2005) Replication of *Yersinia pestis* in interferon g-activated macrophages requires *ripA*, a gene encoded in the pigmentation locus. Proc Natl Acad Sci 102(36):12909–12914
28. Noel B, Lilo S, Capurso D, Hill J, Bliska J (2009) *Yersinia pestis* can bypass protective antibodies to LcrV and activation with gamma interferon to survive and induce apoptosis in murine macrophages. Clin Vaccine Immunol 16(10):1457–1466
29. Ruckdeschel K, Pfaffinger G, Haase R, Sing A, Weighardt H, Hacker G, Holzmann B, Heesemann J (2004) Signaling of apoptosis through TLRs critically involves Toll/IL-1 receptor domain-containing adapter inducing IFNβ but not MyD88, in bacteria-infected murine macrophages. J Immunol 173:3320–3328
30. Straley S, Harmon P (1984) Growth in mouse peritoneal macrophages of *Yersinia pestis* lacking established virulence determinants. Infect Immun 45(3):649–654
31. Pujol C, Bliska J (2003) The ability to replicate in macrophages is conserved between *Yersinia pestis* and *Yersinia pseudotuberculosis*. Infect Immun 71(10):5892–5899
32. Ke Y, Chen Z, Yang R (2013) *Yersinia pestis*: mechanisms of entry into and resistance to the host cell. Front Cell Infect Microbiol 3:106
33. Pujol C, Klein K, Romanov G, Palmer L, Cirota C, Zhao Z, Bliska JB (2009) *Yersinia pestis* can reside in autophagosomes and avoid xenophagy in murine macrophages by preventing vacuole acidification. Infect Immun 77(6):2251–2261
34. Grabenstein J, Marceau M, Pujol C, Simonet M, Bliska J (2004) The response regulator PhoP of *Yersinia pseudotubeculosis* is important for replication in macrophages and for virulence. Infect Immun 72(9):4973–4984
35. Oyston P, Dorrell N, Williams K, Shu-Rui L, Green M, Titball R, Wren B (2000) The response regulator PhoP is important for survival under conditions of macrophage-induced stress and virulence in *Yersinia pestis*. Infect Immun 68(6):3419–3425
36. Furrie E, Macfarlane S, Thomson G, Macfarlane G (2005) Toll-like receptors 2, -3 and -4 expression patterns on human colon and their regulation by mucosal-associated bacteria. Immunology 115(4):565–574
37. Doyle S, Vaidya S, O'Connell R, Dadgostar H, Dempsey P, Wu T, Rao G, Sun R, Haberland M, Modlin R, Cheng G (2002) IRF3 mediates a TLR3/TLR4-specific antiviral gene program. Immunity 17:251–263

38. Telepnev M, Klimpel G, Haithcoat J, Knirel Y, Anisimov A, Motin V (2009) Tetraacylated lipopolysaccharide of *Yersinia pestis* can inhibit multiple Toll-like receptor-mediated signaling pathways in human dendritic cells. J Infect Dis 11:1694–1702
39. Honda K, Ohba Y, Yanai H, Negishi H, Mizutani T, Takaoka A, Taya C, Taniguchi T (2005) Spatiotemporal regulation of MyD88-IRF7 signalling for robust type-I interferon induction. Nature 434:1035–1040
40. Mancuso G, Gambuzza M, Midiri A, Biondo C, Papasergi S, Akira S, Teti G, Beninati C (2009) Bacterial recognition by TLR7 in the lysosomes of conventional dendritic cells. Nat Immunol 10(6):587–594
41. Parker D, Prince A (2011) Type I interferon response to extracellular bacteria in the airway epithelium. Trends Immunol 32(12):582–588
42. Parker D, Prince A (2012) *Staphylococcus aureus* induces type I IFN signaling in dendritic cells via TLR9. J Immunol 189(8):4040–4046
43. Weiss G, Maaetoft-Udsen K, Stifter S, Hertzog P, Goriely S, Thomsen A, Paludan S, Frokiaer H (2012) MyD88 drives IFNβ response to *Lactobacillus acidophilus* in dendritic cells through a mechanism involving IRF1, IRF3, and IRF7. J Immunol 189:2860–2868
44. Hickey A, Lin J, Kummer L, Szaba F, Duso D, Tighe M, Parent M, Smiley S (2013) Intranasal prophylaxis with CpG oligodeoxynucleotide can protect against *Yersinia pestis* infection. Infect Immun 81(6):2123–2132
45. Minnich S, Rohde H (2007) A rationale for repression and/or loss of motility by pathogenic *Yersinia* in the mammalian host. Adv Exp Med Biol 603:298–310
46. Robinson N, McComb S, Mulligan R, Dudani R, Krishnan L, Sad S (2012) Type I interferon induces necroptosis in macrophages during infection with *Salmonella enterica* serovar Typhimurium. Nat Immunol 13(10):954–962
47. Gibbert K, Schlaak J, Yang D, Dittmer U (2013) IFNα subtypes: distinct biological activities in anti-viral therapy. Br J Pharmacol 168(5):1048–1058
48. Schiavoni G, Mattei F, Gabriele L (2013) Type I interferons as stimulators of DC-mediated cross-priming: impact on anti-tumor response. Front Immunol 4(483):1–7

Induction and Function of Type I IFNs During Chlamydial Infection

Uma M. Nagarajan

Introduction

Chlamydia trachomatis infection is the leading sexually transmitted bacterial infection (STI) in the US, as reported by CDC. The global burden of chlamydial infection is likely higher than that reported for STI, as ocular trachoma caused by chlamydiae continues to be the leading cause of preventable blindness in the world [1]. *Chlamydia* spp. also cause significant disease in livestock. In women, *C. trachomatis* is a major cause of pelvic inflammatory disease, ectopic pregnancy, and infertility [2]. Chlamydial infections can be self-limiting, providing evidence for the development of protective immune responses [3, 4]. However, infection induces mostly short-term immunity that is strain (serovar) specific, so the risk of re-infection is high, and carries an increased risk of tissue damaging effects [5]. Human epidemiologic studies also indicate increased risk of disease with repeated infection [6, 7]. Consequently, a great deal of research has focused on understanding chlamydial biology and the immune responses to chlamydial infection, with an obvious goal to develop a vaccine that will induce protective responses to *Chlamydia* while avoiding responses that lead to pathology. In this chapter, we will focus on one such innate immune response, the type I IFNs in chlamydial pathogenesis, with emphasis on their role during infection and the mechanism of induction during chlamydial infection.

U.M. Nagarajan (✉)
Division of Pediatric Infectious Diseases, University of North Carolina at Chapel Hill, 8312 MBRB bldg., Chapel Hill, NC, USA
e-mail: nagaraja@email.unc.edu

© Springer International Publishing Switzerland 2014

D. Parker (ed.), *Bacterial Activation of Type I Interferons*,
DOI 10.1007/978-3-319-09498-4_9

The Pathogen and Pathogenesis

Gram-negative *Chlamydia* sp. are obligate intracellular pathogens with a relatively small genome (1–1.3 Mbp) and a unique developmental cycle [8, 9]. The first step in the intracellular chlamydial infection is attachment of the infectious and metabolically inactive elementary body (EB) to the host cell surface. Once the EB enters the cell by endocytosis, it modifies the vacuole to inhibit phagolysosome fusion, and remains confined in a membrane bound vacuole, termed the "inclusion," during its entire developmental cycle [10–12]. Inside the early inclusion the EB transforms into the metabolically active reticulate body (RB) form by a process that involves DNA de-condensation [13] and reductive cleavage of the outer membrane protein complex [14, 15]. Rather than being strictly non-fusogenic with the host vesicular trafficking pathways, the chlamydial inclusion selectively fuses with sphingomyelin containing exocytic vesicles en route to the plasma membrane from the Golgi [16, 17]. Inclusion formation and acquisition of sphingomyelin are initiated very early in the cycle [17], a phenomenon driven by early chlamydial protein synthesis [18]. Inside the inclusion, the RB multiplies by binary fission [10] and the inclusion expands to occupy significant parts of the cytosol during *C. trachomatis* infection. Specific molecular triggers generated in the RBs likely due to its local environment, initiate the conversion of RBs to EBs towards the latter part of the chlamydial developmental cycle, a process that occurs asynchronously. Eventually, by multiple exit mechanisms [19], the infected cells are lysed and the released EBs go on to infect neighboring cells. The sequential conversion from the specialized EB to RB and then back to EB is a unique feature of chlamydial biology.

Unlike several facultative intracellular pathogens, chlamydiae are not equipped with toxins that damage the host cells. *C. trachomatis* strictly infects mucosal epithelial cells during a genital infection and conjunctival cells during an ocular infection. The host response to *C. trachomatis* is initiated by infected epithelial cells [20] and sustained by professional inflammatory cells and neighboring uninfected cells. Chlamydial ligands recognized by surface and intracellular pathogen recognition receptors (PRRs) initiate chemokine and cytokine production as early as 3 h post infection in vivo, suggesting that entry of viable chlamydiae into host cells is sufficient to induce a response [21]. Using the mouse model of chlamydial genital tract infection [22], it has been shown that inflammatory responses are a major determining factor in oviduct pathology. Following bacterial ascension to the oviducts, infected epithelial cells respond to bacterial signals by producing cytokines and chemokines [20] that act locally to recruit PMNs and other immune cells [21]. PMNs are partially protective in the cervix and uterus because they restrict on-going chlamydial replication, amplify cytokine signaling and reduce pathogen load by attacking infected cells [23]. However, PMN recruitment to the oviducts is excessive and prolonged, leading to distal blockage and formation of hydrosalpinx or salpangitis [21, 23–26]. The contribution of innate immune pathways, such as TLR2, IL-1R, IFNAR, TNFR in PMN recruitment and oviduct pathology has been demonstrated using gene knockout mice [27–33]. On the other hand, CD4[+] Th1 cells that produce a predominant IFN-γ response are critical to the control of chlamydial genital and ocular infection, and enhanced Th1 immune responses correlate with protection from infection and disease in both animal models and humans [34–38].

Induction of Type I IFNs During Chlamydial Infection and Its Biological Implications

Induction of type I IFNs has been observed in multiple cell types including oviduct epithelial cells [39], macrophages [40, 41], fibroblasts (McCoy cells) [42], and mouse DC [43] infected in vitro with multiple chlamydial strains. Therefore, the ability to induce IFN-β in response to intracellular *C. trachomatis* infection appears relatively conserved. *C. pneumoniae*, on the other hand does not induce significant levels of IFN-β expression in epithelial cells, which could be a result of its ability to degrade TRAF3 [44].

Type I IFNs are largely inhibitory to chlamydial growth during in vitro infections. Early studies showed significant inhibition of *C. trachomatis* infectivity in HeLa cells treated with different isotypes of IFN-α [45]. IFN-β treatment of macrophages treated with LPS also resulted in significant killing of *C. psittaci*, as observed with IFN-γ-treatment and this was attributed to activation of indoleamine dioxygenase (IDO) activity. IDO decyclizes tryptophan to *N*-formyl kyneurine resulting in reduction in tryptophan pool in the cells affecting chlamydial growth [46]. Further, inhibition of chlamydial growth by TNFα was shown to be partly mediated through an autocrine function of IFN-β enhancing the activity of IDO and could be blocked by tryptophan [47]. Treatment of murine fibroblasts (L cells) with type I IFNs was also shown to significantly reduce the yield of *C. trachomatis* LGV biovar [48]. Besides its role in chlamydial killing, IFN-β was shown to contribute to IFN-γ expression and in induction of CXCL10 in mouse macrophages infected with *C. pneumoniae* [40] and *C. muridarum* [41], respectively.

The protective effect of IFN-α/β observed in vitro was not recapitulated during in vivo *C. muridarum* infection, both in the lungs and genital tract. In the lung infection model, *Ifnar*$^{-/-}$ mice showed less bacterial burden, weight loss, and less pathology in comparison to control mice, which was attributed to lower macrophage apoptosis in the absence of IFNAR signaling [49]. During genital *C. muridarum* infection, *Ifnar*$^{-/-}$ mice displayed a slightly enhanced clearance of infection and significantly reduced oviduct pathology [28]. The improved bacterial clearance in *Ifnar*$^{-/-}$ mice was associated with an increase in antigen-specific T cells in the iliac nodes, enhanced CD4[+] T cell recruitment to the genital tract and an increased level of the IFN-γ-inducible protein, CXCL9 in genital secretion. A similar outcome of overall enhanced infection clearance and reduced pathology was observed in genital chlamydial infection during IFN-β neutralization in wild-type mice [30]. However, in this study a slight increase in infectious burden was observed at day 4 post infection during IFN-β depletion, which was not sustained and the IFN-β depleted mice went on to clear infection at a faster rate than the mice receiving control sera. A similar outcome of increased infection load at day 4 post infection, which was not sustained was also observed in mice deficient for the transcription factor IRF3, which is essential for IFN-β induction. These data suggest that IFN-β likely has an anti-chlamydial activity early in infection, but its negative impact on the inflammatory cells and T cells is not protective to the host. Indeed, the T cells from the iliac lymph nodes of *Irf3*$^{-/-}$ and *Ifnar*$^{-/-}$ mice displayed an enhanced antigen-specific T cell response. IRF3 KO mice also developed significant uterine pathology unlike *Ifnar*$^{-/-}$ or IFN-β depleted mice, suggesting that IRF3 could play an IFNAR/IFN-β independent role in uterine horn protection [30].

Recent discovery of a new family member of type I IFNs, IFN epsilon (IFN-ε), has generated interest due to its exclusive expression in the mouse and human genital tract [50, 51]. *Ifnε*$^{-/-}$ mice were shown to have slightly enhanced chlamydial infection, suggesting a protective role in infection. IFN-ε signals by the same receptor, IFNAR used by IFN-α/β. Possible explanations for the discordant results during infection between *Ifnε*$^{-/-}$ and *Ifnar*$^{-/-}$ could be a result of a direct role for IFN-ε in chlamydial killing, its regulation by sex hormones and/or a constitutive role in imparting resistance to genital tract epithelia in an IFNAR-independent manner.

To understand the mechanism behind the conflicting role of type I IFN in vitro and in vivo, the pleiotropic immune functions of this cytokine in vivo needs further understanding. Type I IFNs are a potent regulator of adaptive immunity, affecting multiple cell types, including macrophages, lymphocytes, and DCs. IFNα/β induces the expression of several interferon response genes (IRG), which are important for Th1 maturation [52]. Type I IFNs have also been implicated in the generation of cytotoxic T cells and promotion of in vivo T cell proliferation [53] and T cell survival [54]. However, type I IFNs are also known to inhibit IFNγ-induced MHC class II expression [55–57] a function that contradicts its Th1 stimulatory role. Type I IFNs have also been shown to inhibit maturation and activation of mouse Langerhans cells [58]. IFN-β has been reported to augment [59] or downregulate IL-12 and CD40 expression in DC [60]. Further, therapeutic administration of IFN-β in multiple sclerosis patients led to inhibition of IL-12, augmentation of IL-10 production [61] and inhibition of IL-1β production [62]. The paradoxical effect of IFN-β on the expression of Th1-type immune responses partly depends on the timing of DC exposure (during maturation vs. mature) to IFNβ [52]. Type I IFNs are also pro-apoptotic and induce the expression of a number of pro-apoptotic genes, which could play a major role in pathological outcomes during infection. Overall, the detrimental effect of IFN-β during chlamydial infection is a likely result of inhibition of Th1-response, a reduction in IFNγ responsiveness, and induction of an apoptotic response. These results have been largely inferred from gene knockout mice studies and antibody depletion studies. It is possible that the pathological outcome could be different if the mice were treated with recombinant IFN-β. Treatment of mice with recombinant IFN-β has been shown to downregulate IL-1β levels significantly at multiple steps [62]. Since IL-1 signaling is a major player in oviduct pathology during genital chlamydial infection, this may be protective to the oviducts during infection [29]. Therefore, the overall effects of IFN-β during chlamydial infection in vivo is likely determined by its levels in the local tissue, and assigning a beneficial or detrimental role to it would be contextual.

Mechanism of IFN-β Induction During Chlamydial Infection

Multiple host PRRs can induce IFN-β expression during viral or bacterial infection [reviewed in [63]]. Purified *E. coli* LPS is a potent stimulator for TLR4 pathway and routinely used as a positive control for TLR4 activation [64]. However, chlamydial lipopolysaccharide (LPS) has low endotoxic activity [65, 66], which is attributed to the higher hydrophobicity of its lipid A moiety with fatty acids of longer chain length and the presence of non-hydroxylated fatty acids ester-linked to the sugar backbone. Therefore, chlamydiae stimulate TLR4 poorly, although there is one report demonstrating detection of chlamydial LPS by TLR2 [67]. Besides LPS, other bacterial ligands can stimulate TLR4, as in the case of purified hsp60 from *Chlamydia* spp. [68] However, during chlamydial infection, there is limited role for TLR4 in IFN-β induction [41]. Cell invasion and intracellular growth is a prerequisite for IFN response during chlamydial infection. This prerequisite would suggest that intracellular receptors would be preferred over membrane-expressed receptors during infection.

Early studies showed that *C. muridarum*-induced IFN-β is independent of TLR2 and TLR4, and some contribution from MyD88 pathway was suggested [41]. However, no contribution of TLR7 and TLR9 in IFN-β expression was observed and TLR4-MyD88 double knockout macrophages induced similar levels of IFN-β compared to WT macrophages [69]. Further, cytosolic RNA sensing RLR and MAVS pathways were dispensable for this response [69]. This study went on to show that the adaptor for DNA sensing, STING was essential for IFN-β induction during *C. muridarum* infection in both mouse and human epithelial cells [69]. STING was found to localize in close proximity to the inclusion [69]. These data suggested that DNA sensors or chlamydial cyclic di-AMP could be contributing to this response. Indeed, recently it was shown that cyclic di-AMP is produced by *C. trachomatis* EBs [70]. The contribution of second messenger cyclic di AMP in IFN-β expression was shown by infecting HEK293T cells overexpressing STING, and transfected with IFN-β promoter-driven luciferase reporter construct and by using fibroblasts from STING-deficient mice. Recent studies from our laboratory involved screening of multiple DNA sensors during chlamydial infection and a predominant contribution of the DNA sensor cGAS was observed in multiple cell types in response to infection using multiple *C. trachomatis* serovar [71]. The discovery of cGAS as a requirement for IFN-β induction during chlamydial infection suggests that chlamydial DNA is available for sensing on the cytosolic side of the inclusion membrane. In support of this, cGAS was found distinctly localized on the cytosolic side of the chlamydial inclusion membrane and significant co-localization of cGAS and STING was observed after infection.

Besides DNA and cyclic dinucleotide sensing by STING, other receptors have also been shown to contribute to *Chlamydia*-induced IFN-β. During *C. pneumoniae* infection in HUVEC cells, signaling through MAVS was found to be essential for IRF3 activation [72]. MAVS associates with TRAF3, leading to activation of IRF transcription factors and IFN-β expression [73]. It was reported that TLR3 contributes to IFN-β induction in mouse oviduct epithelial cells, during *C. muridarum* infection [74]. The requirement for TLR3 for IFN-β in a bacterial infection is unique to *Chlamydia*. However, it has not been shown how TLR3 interacts with the chlamydial inclusion and the nature of chlamydial ligand engaged is unclear. Since STING was shown to be essential for IFN-β expression during chlamydial infection in the same cell type [69], it is unclear if there is any interaction between the two pathways. Taken together, these data suggest the use of more than one host receptors for IFN-β induction during chlamydial infection. These differences in observation could be due to: (1) differences between *C. trachomatis* vs. *C. pneumoniae* infection, (2) use of multiple pathways in cell types tested and their ability to compensate for each other, and/or (3) infection dose. The use of multiple receptors to induce the expression of the same cytokine may not be unique to *Chlamydia* spp., as multiple receptors have been suggested to play a role in IFN-β induction during *L. monocytogenes* infection [75–78]. The signaling of type I IFN by *Chlamydia* is summarized in Fig. 1.

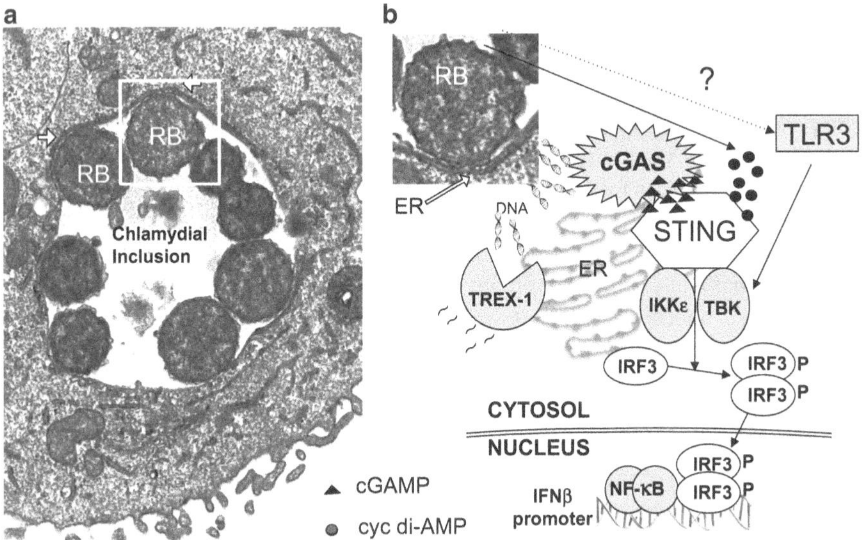

Fig. 1 Model(s) for IFN beta expression during chlamydial infection. (**a**) An electron micrograph of chlamydial inclusion containing metabolically active RBs. (**b**) An enlarged image of an RB, its interaction with the host ER outside the inclusion membrane, and the proposed model(s) for IFNβ expression during infection. At least three models have been proposed for chlamydial recognition with two demonstrating the requirement for the adaptor protein STING in IFNβ induction during infection. In the first model, chlamydial EB (not shown) produce cyc di-AMP that directly interacts with STING to result in IFN-β induction. In the second, the host DNA sensor cGAS was found to be essential for this response, with evidence for cGAMP production during infection. In support of this model, cells lacking the exonuclease TREX-1 show enhanced IFN-β expression during infection, implicating DNA as a ligand for this response. In an alternative third model, TLR3 knock down resulted in a decrease in IFN-β expression in a mouse oviduct cell line. The ligand engaged and its interaction with TLR3 is unknown

Chlamydial Ligands for IFN-β Response

The presence of enzymes essential for cyclic-di AMP synthesis and the demonstration of the presence of cyclic-di AMP in *C. trachomatis* EB [70] indicates cyclic-di AMP as a compelling ligand for IFNβ induction. Simultaneously, the significant requirement for cGAS for *Chlamydia*-induced IFN-β indicates DNA and cGAMP as a possible ligand for IFNβ induction during chlamydial infection [71]. Although this study does not show a direct interaction of DNA with cGAS, evidence for cGAMP production was provided by demonstration of its functional transfer. HeLa cells knocked down for cGAS or STING lose their ability to induce IFN-β upon infection, which was surprisingly rescued following their co-culture. These data suggest that cGAMP produced in cGAS competent cells during infection can function in trans by migrating to STING^{+} cells to induce IFN-β. These results are based on a recent study that showed that cGAMP can cross gap junctions between epithelial cells and provide cells adjacent to an infected cell intrinsic immunity independent of IFNAR signaling [79]. cGAMP binds to the same pocket in STING as cyclic di-AMP/di-GMP, but at a much lower concentration with higher affinity [80]. Indeed, the cGAS product, 2′3′cGAMP, is a much more potent ligand of STING than all other bacterial cyclic di-nucleotides described [81]. Further, human STING is responsive only to cGAMP and unresponsive to the STING ligands CMA [82] and cyclic di-AMP/cyclic di-GMP [83], unlike mouse STING which is responsive to both cyclic dinucleotides and cGAMP [84]. These studies significantly shift the importance of cGAMP over bacterial cyclic dinucleotides during *C. trachomatis* infection in human cells. However, how chlamydial DNA is transferred to cytosol remains unclear at this point. Manzanillo et al. [85] have shown that during *Mycobacterium tuberculosis* infection, phagosomal permeabilization mediated by the bacterial ESX-1 secretion system allows cytosolic recognition pathways access to DNA [85]. Numerous studies have linked IFN β expression to bacterial secretion systems [78, 86–88]. Small molecule inhibitors of type III secretion system (T3SS) were shown to abrogate IFN-β expression in *C. muridarum* infected cells [89], suggesting a similar role for chlamydial T3SS in permeabilization of inclusion membrane. Previous studies [90] have shown that chlamydial reticulate bodies (RB) make direct contact with the inclusion membrane, likely through T3SS. These could be potential permeabilization points where nucleic acids could leak into cytosol and made available for host recognition. It has been shown that *Chlamydia* hijacks the host ER and several ER proteins were found localized on inclusion membrane [91, 92]. The localization of the ER protein STING [69] and cytosolic cGAS in close proximity to the inclusion membrane suggest that STING could serve as a membrane scaffold for the interactions between DNA-cGAS to take place. An alternative source of DNA detected could be host mitochondrial DNA released following damage to mitochondria in *Chlamydia*-infected cells. This argument is countered by the observation that *Chlamydia* spp. inhibit host apoptosis and no mitochondrial damage has been observed in the first 24 h of infection [93]. However, in the environment of other innate receptor recognition and production of cytokines like TNFα, it is possible that mitochondrial damage occurs during in vivo infection and may also contribute to DNA sensor-mediated activation.

Future Directions and Perspectives

A fascinating feature of IFN-β inducing pathways is the resources used by the cell to detect a wide variety of pathogens to generate this important cytokine. During evolution of the immune system, viral infections likely drove this arm of innate immunity to the complex form to which it presently exists. During infection with an intracellular bacteria, the host cells responds as it would to a viral infection, detecting cytoplasmic nucleic acids and producing IFN-β. However, IFN-β is insufficient to eradicate bacterial infection and not protective to the host during in vivo infection, as in the case of chlamydial infection. In such circumstances, one can speculate that the intracellular bacteria likely exploit this antiviral pathway to its advantage. For an STI pathogen such as *C. trachomatis* that does not cause death, this would result in a prolonged infection period in the host leading to increased transmissibility. Based on studies from the mouse model, we can predict that type I IFNs likely contribute to the persistent chlamydial infection reported in humans. This could be particularly relevant during chlamydial-viral co-infection. For instance during co-infection of *C. trachomatis* with human papilloma virus, the type I IFN response resulting from the viral infection is likely to benefit *C. trachomatis* infection. Whether this results in persistent infection for either or both pathogens is not clear, although there is some evidence for *C. trachomatis* infection to be a risk factor for persistent HPV infection [94]. Over the last decade, a lot has been learnt about type I IFN induction and its role in chlamydial infection. However, the exact molecular mechanism involved in IFN-β mediating host pathology is unclear. Further, the interaction of multiple PRRs and their cell type-specific role needs further elucidation. Over the following decade, we expect discovery of antagonists that can potentially block the pathological arm of this innate response during infection, simultaneously enhancing a protective T cell response during chlamydial infection.

References

1. Tabbara KF (2001) Trachoma: a review. J Chemother 13(Suppl 1):18–22
2. Mardh PA (2004) Tubal factor infertility, with special regard to chlamydial salpingitis. Curr Opin Infect Dis 17(1):49–52
3. Parks KS et al (1997) Spontaneous clearance of *Chlamydia trachomatis* infection in untreated patients. Sex Transm Dis 24(4):229–235
4. Morre SA et al (2002) The natural course of asymptomatic *Chlamydia trachomatis* infections: 45% clearance and no development of clinical PID after one-year follow-up. Int J STD AIDS 13(Suppl 2):12–18

5. Woolridge RL et al (1967) Long-term follow-up of the initial (1959–1960) trachoma vaccine field trial on Taiwan. Am J Ophthalmol 63(5)Suppl:1650–1655
6. Hillis SD et al (1997) Recurrent chlamydial infections increase the risks of hospitalization for ectopic pregnancy and pelvic inflammatory disease. Am J Obstet Gynecol 176(1 Pt 1): 103–107
7. Kimani J et al (1996) Risk factors for *Chlamydia trachomatis* pelvic inflammatory disease among sex workers in Nairobi, Kenya. J Infect Dis 173(6):1437–1444
8. Wyrick PB (2000) Intracellular survival by *Chlamydia*. Cell Microbiol 2(4):275–282
9. Bavoil PM, Hsia R, Ojcius DM (2000) Closing in on *Chlamydia* and its intracellular bag of tricks. Microbiology 146(Pt 11):2723–2731
10. Eissenberg LG, Wyrick PB (1981) Inhibition of phagolysosome fusion is localized to *Chlamydia psittaci*-laden vacuoles. Infect Immun 32(2):889–896
11. Heinzen RA et al (1996) Differential interaction with endocytic and exocytic pathways distinguish parasitophorous vacuoles of *Coxiella burnetii* and *Chlamydia trachomatis*. Infect Immun 64(3):796–809
12. Scidmore M, Fischer E, Hackstadt T (1996) Sphingolipids and glycoproteins are differentially trafficked to the *Chlamydia trachomatis* inclusion. J Cell Biol 134(2):363–374
13. Grieshaber NA et al (2004) Chlamydial histone-DNA interactions are disrupted by a metabolite in the methylerythritol phosphate pathway of isoprenoid biosynthesis. Proc Natl Acad Sci U S A 101(19):7451–7456
14. Hatch TP, Allan I, Pearce JH (1984) Structural and polypeptide differences between envelopes of infective and reproductive life cycle forms of *Chlamydia* spp. J Bacteriol 157(1):13–20
15. Hatch TP, Miceli M, Sublett JE (1986) Synthesis of disulfide-bonded outer membrane proteins during the developmental cycle of *Chlamydia psittaci* and *Chlamydia trachomatis*. J Bacteriol 165(2):379–385
16. Hackstadt T, Scidmore MA, Rockey DD (1995) Lipid metabolism in *Chlamydia trachomatis*-infected cells: directed trafficking of Golgi-derived sphingolipids to the chlamydial inclusion. Proc Natl Acad Sci U S A 92(11):4877–4881
17. Hackstadt T et al (1996) *Chlamydia trachomatis* interrupts an exocytic pathway to acquire endogenously synthesized sphingomyelin in transit from the Golgi apparatus to the plasma membrane. EMBO J 15(5):964–977
18. Scidmore MA et al (1996) Vesicular interactions of the *Chlamydia trachomatis* inclusion are determined by chlamydial early protein synthesis rather than route of entry. Infect Immun 64(12):5366–5372
19. Hybiske K, Stephens RS (2007) Mechanisms of host cell exit by the intracellular bacterium *Chlamydia*. Proc Natl Acad Sci U S A 104(27):11430–11435
20. Rasmussen SJ et al (1997) Secretion of proinflammatory cytokines by epithelial cells in response to *Chlamydia* infection suggests a central role for epithelial cells in chlamydial pathogenesis. J Clin Invest 99(1):77–87
21. Rank RG et al (2010) Host chemokine and cytokine response in the endocervix within the first developmental cycle of *Chlamydia muridarum*. Infect Immun 78(1):536–544
22. Barron AL et al (1981) A new animal model for the study of *Chlamydia trachomatis* genital infections: infection of mice with the agent of mouse pneumonitis. J Infect Dis 143(1):63–66
23. Rank RG et al (2008) Chlamydiae and polymorphonuclear leukocytes: unlikely allies in the spread of chlamydial infection. FEMS Immunol Med Microbiol 54(1):104–113
24. Imtiaz MT et al (2007) A role for matrix metalloproteinase-9 in pathogenesis of urogenital *Chlamydia muridarum* infection in mice. Microbes Infect 9(14–15):1561–1566
25. Lee HY et al (2010) A role for CXC chemokine receptor-2 in the pathogenesis of urogenital *Chlamydia muridarum* infection in mice. FEMS Immunol Med Microbiol 60(1):49–56
26. Frazer LC et al (2011) Enhanced neutrophil longevity and recruitment contribute to the severity of oviduct pathology during *Chlamydia muridarum* infection. Infect Immun 79(10):4029–4041
27. Darville T et al (2003) Toll-like receptor-2, but not Toll-like receptor-4, is essential for development of oviduct pathology in chlamydial genital tract infection. J Immunol 171(11): 6187–6197

28. Nagarajan UM et al (2008) Type I interferon signaling exacerbates *Chlamydia muridarum* genital infection in a murine model. Infect Immun 76(10):4642–4648
29. Nagarajan UM et al (2012) Significant role of IL-1 signaling, but limited role of inflammasome activation, in oviduct pathology during *Chlamydia muridarum* genital infection. J Immunol 188(6):2866–2875
30. Prantner D et al (2011) Interferon regulatory transcription factor 3 protects mice from uterine horn pathology during *Chlamydia muridarum* genital infection. Infect Immun 79(10): 3922–3933
31. Prantner D et al (2009) Critical role for interleukin-1beta (IL-1beta) during *Chlamydia muridarum* genital infection and bacterial replication-independent secretion of IL-1beta in mouse macrophages. Infect Immun 77(12):5334–5346
32. Cheng W et al (2008) Caspase-1 contributes to *Chlamydia trachomatis*-induced upper urogenital tract inflammatory pathologies without affecting the course of infection. Infect Immun 76(2):515–522
33. Murthy AK et al (2011) Tumor necrosis factor alpha production from $CD8^+$ T cells mediates oviduct pathological sequelae following primary genital *Chlamydia muridarum* infection. Infect Immun 79(7):2928–2935
34. Rank RG, Soderberg LS, Barron AL (1985) Chronic chlamydial genital infection in congenitally athymic nude mice. Infect Immun 48(3):847–849
35. Rank RG, Barron AL (1983) Effect of antithymocyte serum on the course of chlamydial genital infection in female guinea pigs. Infect Immun 41(2):876–879
36. Rank RG et al (1992) Effect of gamma interferon on resolution of murine chlamydial genital infection. Infect Immun 60(10):4427–4429
37. Cotter TW et al (1997) Dissemination of *Chlamydia trachomatis* chronic genital tract infection in gamma interferon gene knockout mice. Infect Immun 65(6):2145–2152
38. Johansson M et al (1997) Studies in knockout mice reveal that anti-chlamydial protection requires TH1 cells producing IFN-gamma: is this true for humans? Scand J Immunol 46(6): 546–552
39. Johnson RM (2004) Murine oviduct epithelial cell cytokine responses to *Chlamydia muridarum* infection include interleukin-12-p70 secretion. Infect Immun 72(7):3951–3960
40. Rothfuchs AG et al (2001) IFN-alpha beta-dependent, IFN-gamma secretion by bone marrow-derived macrophages controls an intracellular bacterial infection. J Immunol 167(11): 6453–6461
41. Nagarajan UM et al (2005) Chlamydia trachomatis induces expression of IFN-gamma-inducible protein 10 and IFN-beta independent of TLR2 and TLR4, but largely dependent on MyD88. J Immunol 175(1):450–460
42. Devitt A et al (1996) Induction of alpha/beta interferon and dependent nitric oxide synthesis during *Chlamydia trachomatis* infection of McCoy cells in the absence of exogenous cytokine. Infect Immun 64(10):3951–3956
43. Shaw JH et al (2001) Expression of genes encoding Th1 cell-activating cytokines and lymphoid homing chemokines by chlamydia-pulsed dendritic cells correlates with protective immunizing efficacy. Infect Immun 69(7):4667–4672
44. Wolf K, Fields KA (2013) *Chlamydia pneumoniae* impairs the innate immune response in infected epithelial cells by targeting TRAF3. J Immunol 190(4):1695–1701
45. de la Maza LM et al (1985) Interferon-induced inhibition of *Chlamydia trachomatis*: dissociation from antiviral and antiproliferative effects. Infect Immun 47(3):719–722
46. Carlin JM, Weller JB (1995) Potentiation of interferon-mediated inhibition of *Chlamydia* infection by interleukin-1 in human macrophage cultures. Infect Immun 63(5):1870–1875
47. Shemer-Avni Y, Wallach D, Sarov I (1988) Inhibition of *Chlamydia trachomatis* growth by recombinant tumor necrosis factor. Infect Immun 56(9):2503–2506
48. Rothermel CD, Byrne GI, Havell EA (1983) Effect of interferon on the growth of *Chlamydia trachomatis* in mouse fibroblasts (L cells). Infect Immun 39(1):362–370
49. Qiu H et al (2008) Type I IFNs enhance susceptibility to *Chlamydia muridarum* lung infection by enhancing apoptosis of local macrophages. J Immunol 181(3):2092–2102

50. Fung KY et al (2013) Interferon-epsilon protects the female reproductive tract from viral and bacterial infection. Science 339(6123):1088–1092
51. Hermant P et al (2013) IFN-epsilon is constitutively expressed by cells of the reproductive tract and is inefficiently secreted by fibroblasts and cell lines. PLoS One 8(8):e71320
52. Nagai T et al (2003) Timing of IFN-beta exposure during human dendritic cell maturation and naive Th cell stimulation has contrasting effects on Th1 subset generation: a role for IFN-beta-mediated regulation of IL-12 family cytokines and IL-18 in naive Th cell differentiation. J Immunol 171(10):5233–5243
53. Tough DF, Borrow P, Sprent J (1996) Induction of bystander T cell proliferation by viruses and type I interferon in vivo. Science 272(5270):1947–1950
54. Belardelli F et al (2002) Interferon-alpha in tumor immunity and immunotherapy. Cytokine Growth Factor Rev 13(2):119–134
55. Heise MT et al (1998) Murine cytomegalovirus infection inhibits IFN gamma-induced MHC class II expression on macrophages: the role of type I interferon. Virology 241(2):331–344
56. Lu HT et al (1995) Interferon (IFN) beta acts downstream of IFN-gamma-induced class II transactivator messenger RNA accumulation to block major histocompatibility complex class II gene expression and requires the 48-kD DNA-binding protein, ISGF3-gamma. J Exp Med 182(5):1517–1525
57. Devajyothi C et al (1993) Inhibition of interferon-gamma-induced major histocompatibility complex class II gene transcription by interferon-beta and type beta 1 transforming growth factor in human astrocytoma cells. Definition of cis-element. J Biol Chem 268(25): 18794–18800
58. Fujita H et al (2005) Type I interferons inhibit maturation and activation of mouse Langerhans cells. J Invest Dermatol 125(1):126–133
59. Gautier G et al (2005) A type I interferon autocrine-paracrine loop is involved in Toll-like receptor-induced interleukin-12p70 secretion by dendritic cells. J Exp Med 201(9): 1435–1446
60. McRae BL, Beilfuss BA, van Seventer GA (2000) IFN-{beta} differentially regulates CD40-induced cytokine secretion by human dendritic cells. J Immunol 164(1):23–28
61. Byrnes AA, McArthur JC, Karp CL (2002) Interferon-beta therapy for multiple sclerosis induces reciprocal changes in interleukin-12 and interleukin-10 production. Ann Neurol 51(2): 165–174
62. Guarda G et al (2011) Type I interferon inhibits interleukin-1 production and inflammasome activation. Immunity 34(2):213–223
63. Stetson DB, Medzhitov R (2006) Type I interferons in host defense. Immunity 25(3): 373–381
64. Tapping RI et al (2000) Toll-like receptor 4, but not toll-like receptor 2, is a signaling receptor for *Escherichia* and *Salmonella* lipopolysaccharides. J Immunol 165(10):5780–5787
65. Heine H et al (2003) Endotoxic activity and chemical structure of lipopolysaccharides from *Chlamydia trachomatis* serotypes E and L2 and *Chlamydophila psittaci* 6BC. Eur J Biochem 270(3):440–450
66. Rund S et al (1999) Structural analysis of the lipopolysaccharide from *Chlamydia trachomatis* serotype L2. J Biol Chem 274(24):16819–16824
67. Erridge C et al (2004) Lipopolysaccharides of *Bacteroides fragilis*, *Chlamydia trachomatis* and *Pseudomonas aeruginosa* signal via toll-like receptor 2. J Med Microbiol 53(Pt 8):735–740
68. Bulut Y et al (2002) Chlamydial heat shock protein 60 activates macrophages and endothelial cells through Toll-like receptor 4 and MD2 in a MyD88-dependent pathway. J Immunol 168(3):1435–1440
69. Prantner D, Darville T, Nagarajan UM (2010) Stimulator of IFN gene is critical for induction of IFN-beta during *Chlamydia muridarum* infection. J Immunol 184(5):2551–2560
70. Barker JR et al (2013) STING-dependent recognition of cyclic di-AMP mediates type I interferon responses during *Chlamydia trachomatis* infection. mBio 4(3)

71. Zhang Y, Yeruva L, Marinov A, Prantner D, Wyrick PB, Lupashin V, Nagarajan UM (2014) The DNA Sensor, Cyclic GMP-AMP Synthase, Is Essential for Induction of IFN-β during Chlamydia trachomatis Infection. J Immunol Sep 1;193(5):2394–404
72. Buss C et al (2010) Essential role of mitochondrial antiviral signaling, IFN regulatory factor (IRF)3, and IRF7 in *Chlamydophila pneumoniae*-mediated IFN-beta response and control of bacterial replication in human endothelial cells. J Immunol 184(6):3072–3078
73. Saha SK et al (2006) Regulation of antiviral responses by a direct and specific interaction between TRAF3 and Cardif. EMBO J 25(14):3257–3263
74. Derbigny WA et al (2011) The *Chlamydia muridarum*-induced IFN-beta response is TLR3-dependent in murine oviduct epithelial cells. J Immunol 185(11):6689–6697
75. Yang P et al (2010) The cytosolic nucleic acid sensor LRRFIP1 mediates the production of type I interferon via a beta-catenin-dependent pathway. Nat Immunol 11(6):487–494
76. Soulat D et al (2006) Cytoplasmic *Listeria monocytogenes* stimulates IFN-beta synthesis without requiring the adapter protein MAVS. FEBS Lett 580(9):2341–2346
77. Ishikawa H, Ma Z, Barber GN (2009) STING regulates intracellular DNA-mediated, type I interferon-dependent innate immunity. Nature 461(7265):788–792
78. Crimmins GT et al (2008) *Listeria monocytogenes* multidrug resistance transporters activate a cytosolic surveillance pathway of innate immunity. Proc Natl Acad Sci U S A 105(29): 10191–10196
79. Ablasser A et al (2013) Cell intrinsic immunity spreads to bystander cells via the intercellular transfer of cGAMP. Nature 503(7477):530–534
80. Wu J et al (2013) Cyclic GMP-AMP is an endogenous second messenger in innate immune signaling by cytosolic DNA. Science 339(6121):826–830
81. Zhang X et al (2013) Cyclic GMP-AMP containing mixed phosphodiester linkages is an endogenous high-affinity ligand for STING. Mol Cell 51(2):226–235
82. Cavlar T et al (2013) Species-specific detection of the antiviral small-molecule compound CMA by STING. EMBO J 32(10):1440–1450
83. Conlon J et al (2013) Mouse, but not human STING, binds and signals in response to the vascular disrupting agent 5,6-dimethylxanthenone-4-acetic acid. J Immunol 190(10):5216–5225
84. Gao P et al (2013) Structure-function analysis of STING activation by c[G(2′,5′)pA(3′,5′)p] and targeting by antiviral DMXAA. Cell 154(4):748–762
85. Manzanillo PS et al (2012) *Mycobacterium Tuberculosis* activates the DNA-dependent cytosolic surveillance pathway within macrophages. Cell Host Microbe 11(5):469–480
86. Stetson DB, Medzhitov R (2006) Recognition of cytosolic DNA activates an IRF3-dependent innate immune response. Immunity 24(1):93–103
87. Roux CM et al (2007) Brucella requires a functional type IV secretion system to elicit innate immune responses in mice. Cell Microbiol 9(7):1851–1869
88. Stanley SA et al (2007) The type I IFN response to infection with *Mycobacterium tuberculosis* requires ESX-1-mediated secretion and contributes to pathogenesis. J Immunol 178(5): 3143–3152
89. Prantner D, Nagarajan UM (2009) Role for the chlamydial type III secretion apparatus in host cytokine expression. Infect Immun 77(1):76–84
90. Wilson DP et al (2006) Type III secretion, contact-dependent model for the intracellular development of chlamydia. Bull Math Biol 68(1):161–178
91. Dumoux M et al (2012) Chlamydiae assemble a pathogen synapse to hijack the host endoplasmic reticulum. Traffic 13(12):1612–1627
92. Giles DK, Wyrick PB (2008) Trafficking of chlamydial antigens to the endoplasmic reticulum of infected epithelial cells. Microbes Infect 10:1494–1503
93. Fan T et al (1998) Inhibition of apoptosis in chlamydia-infected cells: blockade of mitochondrial cytochrome c release and caspase activation. J Exp Med 187(4):487–496
94. Silins I et al (2005) *Chlamydia trachomatis* infection and persistence of human papillomavirus. Int J Cancer 116(1):110–115

Regulation of Host Response to Mycobacteria by Type I Interferons

Sebastian A. Stifter, Mikaela C. Coleman, and Carl G. Feng

Mycobacterial Infection and Tuberculosis

Mycobacteria are slow growing, facultative intracellular bacilli that primarily reside in macrophages. The *Mycobacterium* genus comprises more than 100 different species. Among them are the pathogenic species *Mycobacterium tuberculosis* and *M. leprae*, causing tuberculosis (TB) and leprosy, respectively. This chapter will discuss mainly on the role of type I interferons (IFNs) in *M. tuberculosis* infection, which is the focus of the majority of recent studies. TB is one of the major infectious diseases worldwide [1]. Two billion people are infected with *M. tuberculosis*, 10 % of whom will eventually develop active TB disease [2]. Annually, more than eight million people develop TB, which is responsible for over 1.3 million deaths, figures still grossly underestimated due to failures in reporting and detection [1, 3]. Once considered to be on its way to extinction, *M. tuberculosis* is posing a significant threat to global health due to the emergence of multi-drug resistant strains [3]. *M. bovis* Bacillus Calmette–Guérin (BCG), the only TB vaccine available for humans, is ineffective in protecting adults against pulmonary TB [4]. Therefore understanding the immune responses to the pathogen may lead to improved vaccination and therapy.

S.A. Stifter • M.C. Coleman • C.G. Feng (✉)
Laboratory of Immunology and Host Defense, Department of Infectious Diseases and Immunology, Sydney Medical School, The University of Sydney, Sydney, NSW 2006, Australia
e-mail: carl.feng@sydney.edu.au

© Springer International Publishing Switzerland 2014

D. Parker (ed.), *Bacterial Activation of Type I Interferons*,
DOI 10.1007/978-3-319-09498-4_10

Host Immune Response to *M. tuberculosis*

Host control of *M. tuberculosis* infection in both humans and mice depends on cell-mediated immunity [5, 6]. Interestingly however, despite the development of an adaptive immune response, some bacilli resist killing and survive within macrophages in granulomas [7, 8]. Mycobacterial granulomas are typically composed of lymphocytes and infected macrophages. The former cell population is thought to provide cytokine mediators necessary for macrophage activation and restriction of intracellular growth of mycobacteria [6, 9]. Importantly, defects in lymphocyte recruitment and retention or effector function during chronic infection can lead to a breakdown of immunity and result in progressive infection [10, 11].

Resistance to mycobacteria is critically dependent on the T helper 1 (Th1) response [5, 6]. Thus, patients and animals deficient in IFN-γ, IL-12, STAT1, or T-cells show significantly increased susceptibility to mycobacterial infections [12]. In addition to $CD4^+$ lymphocytes, natural killer (NK) cells also contribute to cytokine production in *M. tuberculosis* infection [13]. Critically, host control of *M. tuberculosis* infection requires intact IFN-γ receptor signaling in both hematopoietic and non-hematopoietic components [14]. It is believed that the key effector function of IFN-γ is to activate infected macrophages to produce antimicrobial mediators, such as nitric oxide and p47 immunity-related GTPases [15]. But, emerging evidence from recent studies indicates that this long-held concept may represent an over-simplified view. For example, in addition to impaired bacterial control, *M. tuberculosis*-infected *Ifng*$^{-/-}$ mice show severe pulmonary pathology associated with dramatically increased accumulation of neutrophils [13, 16]. Together, these findings suggest that IFN-γ plays a broader role in inflammation and infection beyond its proposed function in bacterial killing.

Mycobacterial Virulence Mechanisms

Following infection, avirulent mycobacteria are effectively cleared by host defence mechanisms and are unable to persist in the host. In contrast, virulent mycobacteria establish persistent infection in the infected host. The ability of *M. tuberculosis* to avoid host antimicrobial strategies is well documented [17–21]. One strategy involves the action of the *M. tuberculosis* mannose-capped lipoarabinomannan that is incorporated into lipid rafts of the plasma membrane, thereby executing arrest in phagosomal maturation [22–25]. The bacterium is known to block fusion of lysosomes as well as inhibit phagosomal acidification [26]. This prevents the activation of a number of pH-dependent antimicrobial compounds, such as maturation of cathepsin D [22], which are required for destroying intracellular bacteria.

It has been assumed that because *M. tuberculosis* is broadly equipped to combat phagosomal maturation via inhibition of Golgi-trafficking, phagosome acidification and lysosomal fusion, that an operational phagosome would be effective in clearing infection [27]. However, the identification of bacterial strategies that allow the survival of *M. tuberculosis* mutants in fully mature phagosomes challenges this assumption. These strategies include the ability to deactivate reactive oxygen species (ROS) and protect against NOS2 damage [8]. Thus, it is unlikely that *M. tuberculosis* causes definitive arrest of phagosomal maturation, but rather delays it [28]. This may be a temporary measure to allow the bacteria to adapt and initiate transcription in response to the intracellular environment [28]. Recent studies suggest that rather than using the mycobacterial phagosome as a replicative niche, as traditionally believed, the mycobacterially-altered phagosomes act as a preparation and "waiting room" for escape of the bacteria into the cytosol [29]. This suggestion has significant ramifications, both for bacterial survival and the host defence mechanisms involved.

In a seminal study, van der Wel et al. [30] report that virulent *M. tuberculosis* but not heat-killed mycobacteria or vaccine strain *M. bovis* BCG are present within the cytosol of macrophages 2 days after infection. This translocation is dependent upon the bacterial secretion system, early secretory antigenic target 6 system 1 (ESX-1) apparatus transcribed by the region of difference-1 (RD-1) genes that are present only in virulent mycobacteria including *M. tuberculosis* [31, 32]. However, interestingly, permeabilization of the phagosomal membrane and cytosolic access to bacterial pathogen-associated molecular patterns (PAMPs) may occur within hours of infection under certain conditions [28], long before complete translocation of bacilli into the cytosol occurs. This may allow bacterial components access to the cytosol and avoid their sensing by endosomal Toll-like receptors (TLRs) [33]. Indeed, *M. tuberculosis* β-lactamase catalytic activity occurs in the cytosol progressively, from less than 2 days post infection [29]. These observations suggest that release of bacterial products precedes complete escape of the pathogen. Regardless of the sequence of events, this "phagosomal escape" represents a newly characterized virulence mechanism of *M. tuberculosis*, although whether RD-1 mediates this purported partial permeabilization of the membrane [34] or allows complete translocation of the bacilli into the cytosol [31] remains unclear. Finally, this phagosomal-cytosolic access hypothesis is further supported by studies demonstrating the ability of virulent (RD-1 competent) *M. tuberculosis* to prime $CD8^+$ T cell responses [35] and activate the inflammasome [36, 37] since both processes require the access of microbial products to cytosolic immune pathways.

Type I IFN Production and IFN Signature

In the case of mycobacterial infections, type I IFNs are produced in vitro by *M. tuberculosis*-infected murine macrophages [38, 39], as well as human monocyte derived macrophages [17], dendritic cells [40], and differentiated monocytic THP-1 cells [41]. Importantly type I IFNs are induced during infection with virulent but not avirulent mycobacteria such as with *M. bovis* BCG [17, 21, 42], indicating that type I IFN induction is unique to virulent mycobacteria.

In mice, type I IFNs and their inducible genes are detected in *M. tuberculosis*-infected tissues [21, 43]. Similarly, in humans IFN-β is expressed in leprosy skin lesions [44]. Although type I IFN cytokine genes are undetectable in peripheral blood of infected human subjects, a large set of IFN-inducible genes are readily detected in the same cells. In a seminal study, Berry et al. [45] identified the presence of a 393 transcript gene signature in whole blood of active TB patients. Further analysis revealed that 86-transcripts can distinguish active TB from other types of inflammatory conditions, such as systemic lupus erythematosus (SLE), which is known to be associated with an enhanced type I IFN gene signature [46–48]. Interestingly, 10–25 % of latently infected subjects also displayed the IFN signature [45]. Considering that approximately 10 % of latent *M. tuberculosis*-infected individuals will eventually develop active TB in their lifetime, it would be interesting to examine whether the IFN gene signature detected in latently infected individuals could be predictive for disease progression.

This type I IFN-induced gene signature, which includes the transcription factors IRF1, IRF7, Oct-1, and proteins of the STAT family, has been confirmed in patients with active TB in multiple recent reports [49–52]. Importantly, these transcriptional profiles are also observed in experimental models of *M. tuberculosis* infection in mice and in vitro in human cell lines [49, 50], providing an avenue to investigate these observations in more detail at a functional level. A study examining the expression of interferon regulated genes (IRGs) in cattle infected with *M. bovis* (the causative agent of bovine TB) found an increase in type I and type II IFN regulated genes such as *CXCL10*, *STAT1*, *IFI16*, *IRF7*, and *OAS1* [53], suggesting that similar virulence mechanisms may be conserved across mycobacterial species.

Leprosy is another major human mycobacterial disease in which a type I IFN gene signature is associated with severe disease outcome. Leprosy has traditionally been classified into two major types; tuberculoid and lepromatous. Historically, the self-healing tuberculoid leprosy is believed to be associated with Th1 responses whereas disseminated lepromatous leprosy, which is characterized by uncontrolled infection and increased tissue pathology, is driven by a strong Th2 response. Interestingly, a recent study demonstrated that lepromatous leprosy is associated with an IFN-β-dependent gene signature [44], providing a novel mechanism for uncontrolled bacterial growth associated with this severe form of the disease. Together, studies in both *M. tuberculosis* and *M. leprae* infections have clearly established that the presence of an IFN-inducible gene signature is associated with disease progression and more severe clinical manifestations, although the issue as to whether the observed gene signature is driven by type I, type II, or both remains unclear.

One of the major challenges in TB management is to effectively distinguish latent infection from active disease and to monitor treatment efficacy. To date, no laboratory test is available for these purposes. Therefore, the identification of the whole blood IFN signature associated with active TB disease has generated considerable interest in the clinical and basic TB research communities because of its potential in diagnosing and monitoring TB disease progression [54, 55]. However, the practical impact of the discovery remains unclear, as multiple recent studies have demonstrated that the IFN-inducible gene signature is also associated with other diseases. For example, melioidosis, a disease caused by the intracellular bacterium *Burkholderia pseudomallei*, also contains the 86-gene IFN signature [52]. Another example is sarcoidosis, a lung disease associated with pulmonary lesions similar to TB, in which the blood IFN gene signature has 80 % overlap with that of active TB patients enrolled in the same study [51] and in the study published by Berry et al. These data suggest that while the IFN signature is associated with acute inflammatory conditions it is not specific for mycobacterial infection and TB.

Regulation of Host Resistance to Mycobacterial Infection by Type I IFN Signaling

The hypervirulence of clinical isolates of *M. tuberculosis* correlates with the enhanced synthesis of endogenous type I IFN. Mice deficient in the receptor for these cytokines display significantly reduced bacterial loads when chronically infected with *M. tuberculosis* [21, 38, 56]. More recently, mice deficient in interferon regulatory factor 3 (IRF3), a key molecule required for type I IFN production, were shown to display significantly increased survival and decreased bacterial load relative to their wild-type counterparts following *M. tuberculosis* infection [34]. Interestingly, in the absence of IFN-γ signaling, type I IFNs play a protective role in *M. tuberculosis* infection as mice deficient in both *Ifngr* and *Ifnar* display increased pathology and mortality than *Ifngr* single deficient mice [57]. Therefore, it appears the detrimental effect of type I IFN is dependent on the presence of an IFN-γ response to *M. tuberculosis*.

Conversely, *M. tuberculosis*-infected mice with elevated type I IFN levels show reduced survival and increased bacillary burden compared to control mice. Intranasal administration of poly:ICLC, a compound that stimulates high-level production of type I IFN, exacerbates pulmonary TB in wild-type but not in type I IFN receptor deficient mice [58]. In addition, increased mycobacterial burden is reported in a mouse model of influenza A virus and *M. tuberculosis* co-infection [59]. Importantly, the impaired host resistance is dependent on type I IFN signaling, suggesting that concurrent viral infection can exacerbate TB by triggering type I IFN production.

Compared to animal studies, the effect of type I IFN on resistance to mycobacterial infection in humans is less clear. Increased incidence of TB disease has been reported in patients with active autoimmune diseases or on treatment with recombinant IFN-α. For example, it is established that type I IFNs play a key role in disease pathogenesis of SLE and an IFN-inducible gene signature is observed in patients with severe disease [46–48, 60, 61]. In this case, increased incidence of TB disease in SLE patients has been reported [62–64]. However, the causal role for SLE in TB disease progression is difficult to establish, as SLE patients are also more susceptible to other bacterial infections [65, 66]. This may be a consequence of the immunosuppressant therapy administered to SLE patients, rather than a direct exacerbation of TB by type I IFN signaling. It would be informative to examine whether TB/SLE patients displaying the IFN gene signature can benefit from combination therapy with the IFN-α blocking antibody currently being evaluated in clinical trials [67].

Type I IFN therapy is used frequently for the treatment of multiple sclerosis, hepatitis C virus, and a number of cancers [68]. The effect of type I IFN therapy on *M. tuberculosis* infection remains controversial, since both beneficial and detrimental effects have been reported [69–74]. In a number of cases administration of IFN-α alone or in combination with standard anti-mycobacterial regimen enhanced mycobacterial control in TB patients. In one study, aerosolized IFN-α lead to earlier resolution of infection and was associated with lower levels of IL-1β, IL-6, and TNF-α in bronchoalveolar lavage fluid [75]. In a second case, inclusion of IFN-β to a four drug regimen of mycobacterial antibiotics resulted in rapid resolution of a previously difficult-to-treat infection [76]. Together, these conflicting clinical data make it difficult to conclude the effect of type I IFNs on *M. tuberculosis* infection and TB disease in humans, although it is plausible that type I IFNs are detrimental to *M. tuberculosis* control in certain circumstances but beneficial in others.

Innate Sensing for Type I IFN Induction

Mycobacteria are complex microorganisms that primarily interact with cells of the phagocyte lineage (macrophages and dendritic cells). Recognition of mycobacterial products by surface and cytosolic pathogen recognition receptors (PRRs) leads to the activation of multiple innate pathways.

Cell surface sensing of *M. tuberculosis* 19 kDa lipoprotein occurs through TLR2 in association with TLR1 or TLR6 [77, 78]. However, these TLRs lack TRIF signaling adaptors and are unable to transduce signals culminating in IRF3 activation and IFN production [79]. In infected DCs, sensing of mycobacterial unmethylated CpG-DNA by TLR9 in the endosomal compartment is inhibited by the mycobacteria [80]. Consequently, TLR signaling is not necessary for *M. tuberculosis-mediated* induction of type I IFNs [19, 21, 42, 81, 82].

Virulent mycobacteria capable of causing damage to the phagosomal membrane can trigger type I IFN production in infected macrophages, suggesting that cytosolic sensing could be responsible for mycobacterium-induced type I IFN production. Damage to the phagosomal membrane is accomplished by the ESX-1 secretion system [21]. Known cytosolic receptors for muropeptides, components of the bacterial cell wall, are the nucleotide-binding oligomerization domain-containing protein (NOD) receptors [42, 43, 83]. In *M. tuberculosis* infection, bacterially derived *N*-glycolyl-muramyl dipeptides (MDPs) in the cytosol have been reported to activate the NOD2 sensor, which leads to the induction of type I IFN production in macrophages [42, 84] along with other functions in the innate and adaptive response [85]. In this model, NOD2 recognition of *M. tuberculosis* MDP triggers the ubiquitination of the receptor-interacting protein kinase 2 (RIP2), which subsequently activates TANK binding kinase 1 (TBK1) to stimulate IRF5 leading to the transcription of type I IFN genes (Fig. 1).

However, a distinct sensing mechanism has been proposed recently. Cox and colleagues have demonstrated that induction of type I IFN genes in murine cells

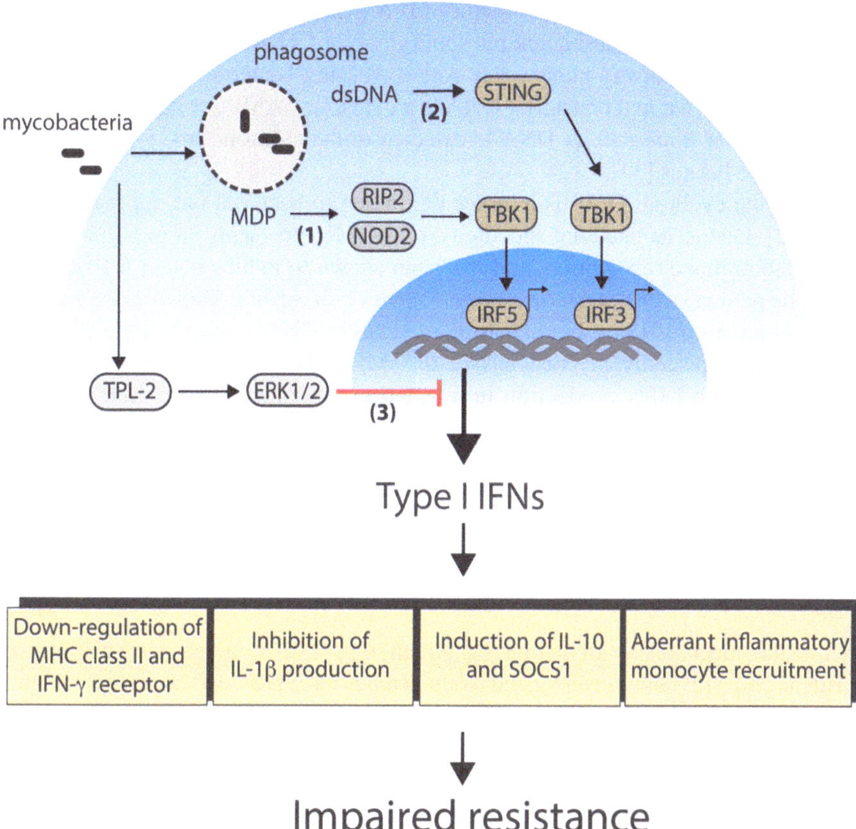

Fig. 1 The postulated mechanisms underlying the induction and function of type I IFN in mycobacterial infection. Phagosomal damage caused by ESX-1, a mycobacterial secretion system present only in virulent mycobacteria, and subsequent cytosolic translocation of microbial products initiates the innate cascades for type I IFN production in mycobacterial-infected macrophages. Although both avirulent and virulent mycobacteria can activate NF-κB (not depicted) and MAPK pathways through surface expressed pattern recognition receptors, only virulent mycobacteria activate cytosolic innate mechanisms leading to type I IFN production. It is postulated that mycobacterial MDP recognized by NOD2 triggers a TBK1 and IRF5-dependent pathway **(1)** whereas mycobacterial double-stranded DNA activates a STING, TBK1, and IRF3-dependent mechanism **(2)**. Both pathways are not effectively triggered by BCG or other RD1-deficient mycobacteria, which lack the ESX-1 secretion system. Finally, activation of the TPL-2/ERK pathway is capable of limiting *M. tuberculosis* induced type I IFN production **(3)** (indicated in *red line*). The production of type I IFN has been associated with increased susceptibility to mycobacterial infection. The cytokines inhibit known antimicrobial effector mechanisms, such as IFN-γ-induced MHC class II upregulation, IFN-γ receptor expression and IL-1β production. Type I IFNs also induce IL-10 and SOCS1 to suppress host protective Th1 response to the pathogen. Finally, over-production of type I IFN exacerbates *M. tuberculosis* infection by recruiting immature myeloid cells that are incompetent in killing intracellular bacteria

requires the activation of the Stimulator of IFN genes (STING) by cytosolic mycobacterial products and subsequent phosphorylation of TBK1 and IRF3 [34]. Due to the fact that the pathway plays a key role in sensing bacterial DNA [86–89], the finding implies that mycobacterial DNA is a cytosolic PAMP for type I IFN induction. Indeed, *M. tuberculosis* DNA is detected in the cytosolic fraction of infected macrophage lysates [34].

Cytosolic cyclic-di-GMP is a molecule unique to bacterial but not mammalian cells [82]. Due to its bacterial specificity, c-di-GMP represents an important target for innate immune recognition, and has been shown to induce potent activation of cytosolic pathways [90]. *Listeria monocytogenes* c-di-AMP is known to induce type I IFN production [91] by binding to the cytosolic DNA sensor STING [92, 93]. However, unexpectedly, mycobacterial c-di-GMP and c-di-AMP are not involved in stimulating type I IFN production in infected murine macrophages [34] although mutations in c-di-GMP signaling within mycobacteria has been linked to impaired infectivity of *M. tuberculosis* [94, 95].

Function of Type I IFN in Mycobacterial Infection

While the mechanisms of action by which type I IFN increase mycobacterial virulence are still being investigated, some studies have provided important insights as to how the cytokines negatively affect anti-mycobacterial immunity (summarized in Fig. 1).

Recent work has demonstrated that type I IFNs regulate the production of IL-1β, a critical cytokine in host resistance to *M. tuberculosis* infection. Novikov et al. [17] demonstrate that exogenous and *M. tuberculosis*-induced type I IFNs are able to suppress *IL1B* gene transcription in human macrophages. Although the exact molecular mechanism responsible for the suppression has yet to be defined, a separate study found that the type I IFN-dependent IL-1β inhibition could be partially restored by blocking IL-10 activity [96]. In addition, IFN-β-dependent IL-10 suppresses IFN-γ-dependent antimicrobial mechanisms in *M. leprae*-infected human macrophages and is associated with the development of lepromatous leprosy [44]. IL-10 is an anti-inflammatory cytokine known to inhibit Th1 responses and macrophage antimicrobial effector functions. Importantly, the cytokine has been shown to exacerbate murine mycobacterial infection in some settings [97–99]. It is, therefore, possible that IL-10 induction by type I IFNs is one of the general mechanisms contributing to the pro-bacterial effect of type I IFNs.

In addition to inducing IL-10, type I IFNs upregulate negative regulators of IFN signaling including suppressor of cytokine signaling 1 (SOCS1) [100, 101]. *Socs1* gene expression is elevated in infection with *M. tuberculosis* strains, particularly those of high virulence and associated with high IFN-α/β stimulating activity [39]. SOCS molecules are also found to be increased in active human TB cases and appear to correlate with disease severity [102, 103]. Experimentally, mouse macrophages deficient in type I IFN receptor demonstrated reduced *Socs1* gene induction after mycobacterial infection [104]. *Socs1* deficient macrophages displayed reduced bacterial numbers compared to wild-type macrophages and, importantly, this effect was dependent on the inhibition of IFN-γ signaling thereby providing a mechanism by which type I IFNs suppress the antimicrobial activities of IFN-γ [104]. In addition, it is well established that type I IFNs can down-regulate the expression of IFN-γ receptor [105, 106]. This down-regulation has also been observed in *M. tuberculosis* infection when infected mice were treated with the synthetic type I IFN inducer poly:ICLC [58]. Limiting the expression of cell surface IFN-γ receptor would likely impair IFN-γ-dependent effector functions in macrophages, such as induction of NOS2.

While multiple mechanisms have been postulated to explain the pro-bacterial function of type I IFNs in mycobacterial infection in vitro, the exact function of the cytokines in vivo is poorly understood. Type I IFN has been shown to down-regulate the Th1 response in one study [38] but not others [56, 58]. Interestingly, the detrimental effect of poly:ICLC treatment in *M. tuberculosis*-infected mice is linked to the dramatically increased accumulation of immature inflammatory monocytes [107]. Flow cytometric sorting experiments reveal that immature myeloid cells harbor significantly more bacteria than their mature counterparts. Therefore, type I IFNs may contribute to host susceptibility to *M. tuberculosis* infection by supplying a niche for mycobacterial growth and survival. The mechanisms that regulate type I IFN production in vivo are also not well defined. However, a recent study demonstrated that mice deficient in tumor progression locus-2 (TPL-2) show increased levels of IFN-β in the serum and bacillary loads in lungs compared to wild-type controls following infection with *L. monocytogenes* and *M. tuberculosis*. It is postulated that the activation of TPL-2 and extracellular signal-regulated kinase 1/2 (ERK1/2) pathways, possibly by TLR signals, prevents excessive production of type I IFN during the infections [108].

Concluding Remarks

Increasing evidence suggests that type I IFNs negatively regulate host resistance to intracellular pathogens including mycobacteria. There is an urgent need for a better understanding of the cytokines' biological functions in infection, as well as identification of the host and mycobacterial factors required for the cytokine production. While the question as to whether an IFN-inducible gene signature will assist in identifying TB cases and latently infected individuals who are at high risk of developing active disease needs to be carefully examined in longitudinal studies, it is clear that type I IFN inducing agents should be used with caution in people with mycobacterial infection. Finally, it would be interesting to examine whether therapeutic blockade of some components of the type I IFN signaling pathway could promote bacterial clearance and reduce TB reactivation and transmission in humans.

References

1. World Health Organisation (2012) Global tuberculosis report 2012. WHO Press, Geneva
2. Zumla A, Raviglione M, Hafner R, von Reyn CF (2013) Tuberculosis. N Engl J Med 368:745–755
3. Dias HM, Falzon D, Fitzpatrick C, Floyd K, Glaziou P, Hiatt T, Lienhardt C, Nguyen L, Sismanidis C, Timimi H, Uplekar M, van Gemert W, Zignol M (2012) Global tuberculosis report. World Health Organisation, Geneva
4. Kaufmann SH, Hussey G, Lambert PH (2010) New vaccines for tuberculosis. Lancet 375: 2110–2119
5. Cooper AM (2009) Cell-mediated immune responses in tuberculosis. Annu Rev Immunol 27:393–422
6. Philips JA, Ernst JD (2012) Tuberculosis pathogenesis and immunity. Annu Rev Pathol 7: 353–384
7. Cosma CL, Sherman DR, Ramakrishnan L (2003) The secret lives of the pathogenic mycobacteria. Annu Rev Microbiol 57:641–676
8. Monack DM, Mueller A, Falkow S (2004) Persistent bacterial infections: the interface of the pathogen and the host immune system. Nat Rev Microbiol 2:747–765
9. Saunders BM, Britton WJ (2007) Life and death in the granuloma: immunopathology of tuberculosis. Immunol Cell Biol 85:103–111
10. Feng CG, Jankovic D, Kullberg M, Cheever A, Scanga CA, Hieny S, Caspar P, Yap GS, Sher A (2005) Maintenance of pulmonary Th1 effector function in chronic tuberculosis requires persistent IL-12 production. J Immunol 174:4185–4192
11. Feng CG, Britton WJ, Palendira U, Groat NL, Briscoe H, Bean AG (2000) Up-regulation of VCAM-1 and differential expansion of beta integrin-expressing T lymphocytes are associated with immunity to pulmonary *Mycobacterium tuberculosis* infection. J Immunol 164:4853–4860
12. Rosenzweig SD, Holland SM (2005) Defects in the interferon-gamma and interleukin-12 pathways. Immunol Rev 203:38–47
13. Feng CG, Kaviratne M, Rothfuchs AG, Cheever A, Hieny S, Young HA, Wynn TA, Sher A (2006) NK cell-derived IFN-gamma differentially regulates innate resistance and neutrophil response in T cell-deficient hosts infected with *Mycobacterium tuberculosis*. J Immunol 177:7086–7093
14. Desvignes L, Ernst JD (2009) Interferon-gamma-responsive nonhematopoietic cells regulate the immune response to *Mycobacterium tuberculosis*. Immunity 31:974–985
15. Taylor GA, Feng CG, Sher A (2007) Control of IFN-γ-mediated host resistance to intracellular pathogens by immunity-related GTPases (p47 GTPases). Microbes Infect 9:1644–1651
16. Nandi B, Behar SM (2011) Regulation of neutrophils by interferon-gamma limits lung inflammation during tuberculosis infection. J Exp Med 208:2251–2262
17. Novikov A, Cardone M, Thompson R, Shenderov K, Kirschman KD, Mayer-Barber KD, Myers TG, Rabin RL, Trinchieri G, Sher A, Feng CG (2011) *Mycobacterium tuberculosis* triggers host type I IFN signaling to regulate IL-1beta production in human macrophages. J Immunol 187:2540–2547
18. Koo IC, Wang C, Raghavan S, Morisaki JH, Cox JS, Brown EJ (2008) ESX-1-dependent cytolysis in lysosome secretion and inflammasome activation during mycobacterial infection. Cell Microbiol 10:1866–1878
19. Shi R, Otomo K, Yamada H, Tatsumi T, Sugawara I (2006) Temperature-mediated heteroduplex analysis for the detection of drug-resistant gene mutations in clinical isolates of *Mycobacterium tuberculosis* by denaturing HPLC, SURVEYOR nuclease. Microbes Infect 8:128–135
20. Kurenuma T, Kawamura I, Hara H, Uchiyama R, Daim S, Dewamitta SR, Sakai S, Tsuchiya K, Nomura T, Mitsuyama M (2009) The RD1 locus in the *Mycobacterium tuberculosis*

genome contributes to activation of caspase-1 via induction of potassium ion efflux in infected macrophages. Infect Immun 77:3992–4001
21. Stanley SA, Johndrow JE, Manzanillo P, Cox JS (2007) The type I IFN response to infection with *Mycobacterium tuberculosis* requires ESX-1-mediated secretion and contributes to pathogenesis. J Immunol 178:3143–3152
22. Fratti RA, Chua J, Vergne I, Deretic V (2003) *Mycobacterium tuberculosis* glycosylated phosphatidylinositol causes phagosome maturation arrest. Proc Natl Acad Sci U S A 100:5437–5442
23. Vergne I, Chua J, Deretic V (2003) Tuberculosis toxin blocking phagosome maturation inhibits a novel Ca2+/calmodulin-PI3K hVPS34 cascade. J Exp Med 198:653–659
24. Kang PB, Azad AK, Torrelles JB, Kaufman TM, Beharka A, Tibesar E, DesJardin LE, Schlesinger LS (2005) The human macrophage mannose receptor directs *Mycobacterium tuberculosis* lipoarabinomannan-mediated phagosome biogenesis. J Exp Med 202:987–999
25. Welin A, Winberg ME, Abdalla H, Sarndahl E, Rasmusson B, Stendahl O, Lerm M (2008) Incorporation of *Mycobacterium tuberculosis* lipoarabinomannan into macrophage membrane rafts is a prerequisite for the phagosomal maturation block. Infect Immun 76: 2882–2887
26. Loeuillet C, Martinon F, Perez C, Munoz M, Thome M, Meylan PR (2006) *Mycobacterium tuberculosis* subverts innate immunity to evade specific effectors. J Immunol 177: 6245–6255
27. van Crevel R, Ottenhoff TH, van der Meer JW (2002) Innate immunity to *Mycobacterium tuberculosis*. Clin Microbiol Rev 15:294–309
28. Welin A, Lerm M (2012) Inside or outside the phagosome? The controversy of the intracellular localization of *Mycobacterium tuberculosis*. Tuberculosis (Edinb) 92:113–120
29. Simeone R, Bobard A, Lippmann J, Bitter W, Majlessi L, Brosch R, Enninga J (2012) Phagosomal rupture by *Mycobacterium tuberculosis* results in toxicity and host cell death. PLoS Pathog 8:e1002507
30. van der Wel N, Hava D, Houben D, Fluitsma D, van Zon M, Pierson J, Brenner M, Peters PJ (2007) *M. tuberculosis* and *M. leprae* translocate from the phagolysosome to the cytosol in myeloid cells. Cell 129:1287–1298
31. Houben D, Demangel C, van Ingen J, Perez J, Baldeon L, Abdallah AM, Caleechurn L, Bottai D, van Zon M, de Punder K, van der Laan T, Kant A, Bossers-de Vries R, Willemsen P, Bitter W, van Soolingen D, Brosch R, van der Wel N, Peters PJ (2012) ESX-1-mediated translocation to the cytosol controls virulence of mycobacteria. Cell Microbiol 14: 1287–1298
32. Abdallah AM, Bestebroer J, Savage ND, de Punder K, van Zon M, Wilson L, Korbee CJ, van der Sar AM, Ottenhoff TH, van der Wel NN, Bitter W, Peters PJ (2011) Mycobacterial secretion systems ESX-1 and ESX-5 play distinct roles in host cell death and inflammasome activation. J Immunol 187:4744–4753
33. Burdette DL, Vance RE (2013) STING and the innate immune response to nucleic acids in the cytosol. Nat Immunol 14:19–26
34. Manzanillo PS, Shiloh MU, Portnoy DA, Cox JS (2012) *Mycobacterium tuberculosis* activates the DNA-dependent cytosolic surveillance pathway within macrophages. Cell Host Microbe 11:469–480
35. Winau F, Weber S, Sad S, de Diego J, Hoops SL, Breiden B, Sandhoff K, Brinkmann V, Kaufmann SH, Schaible UE (2006) Apoptotic vesicles crossprime CD8 T cells and protect against tuberculosis. Immunity 24:105–117
36. Saiga H, Kitada S, Shimada Y, Kamiyama N, Okuyama M, Makino M, Yamamoto M, Takeda K (2012) Critical role of AIM2 in *Mycobacterium tuberculosis* infection. Int Immunol 24:637–644
37. Yang Y, Zhou X, Kouadir M, Shi F, Ding T, Liu C, Liu J, Wang M, Yang L, Yin X, Zhao D (2013) The AIM2 inflammasome is involved in macrophage activation during infection with virulent *Mycobacterium bovis* strain. J Infect Dis 208:1849–1858

38. Manca C, Tsenova L, Bergtold A, Freeman S, Tovey M, Musser JM, Barry CE 3rd, Freedman VH, Kaplan G (2001) Virulence of a *Mycobacterium tuberculosis* clinical isolate in mice is determined by failure to induce Th1 type immunity and is associated with induction of IFN-alpha/beta. Proc Natl Acad Sci U S A 98:5752–5757
39. Manca C, Tsenova L, Freeman S, Barczak AK, Tovey M, Murray PJ, Barry C, Kaplan G (2005) Hypervirulent *M. tuberculosis* W/Beijing strains upregulate type I IFNs and increase expression of negative regulators of the Jak-Stat pathway. J Interferon Cytokine Res 25:694–701
40. Remoli ME, Giacomini E, Lutfalla G, Dondi E, Orefici G, Battistini A, Uze G, Pellegrini S, Coccia EM (2002) Selective expression of type I IFN genes in human dendritic cells infected with *Mycobacterium tuberculosis*. J Immunol 169:366–374
41. Weiden M, Tanaka N, Qiao Y, Zhao BY, Honda Y, Nakata K, Canova A, Levy DE, Rom WN, Pine R (2000) Differentiation of monocytes to macrophages switches the *Mycobacterium tuberculosis* effect on HIV-1 replication from stimulation to inhibition: modulation of interferon response and CCAAT/enhancer binding protein beta expression. J Immunol 165: 2028–2039
42. Pandey AK, Yang Y, Jiang Z, Fortune SM, Coulombe F, Behr MA, Fitzgerald KA, Sassetti CM, Kelliher MA (2009) NOD2, RIP2 and IRF5 play a critical role in the type I interferon response to *Mycobacterium tuberculosis*. PLoS Pathog 5:e1000500
43. Trinchieri G (2010) Type I interferon: friend or foe? J Exp Med 207:2053–2063
44. Teles RM, Graeber TG, Krutzik SR, Montoya D, Schenk M, Lee DJ, Komisopoulou E, Kelly-Scumpia K, Chun R, Iyer SS, Sarno EN, Rea TH, Hewison M, Adams JS, Popper SJ, Relman DA, Stenger S, Bloom BR, Cheng G, Modlin RL (2013) Type I interferon suppresses type II interferon-triggered human anti-mycobacterial responses. Science 339:1448–1453
45. Berry MP, Graham CM, McNab FW, Xu Z, Bloch SA, Oni T, Wilkinson KA, Banchereau R, Skinner J, Wilkinson RJ, Quinn C, Blankenship D, Dhawan R, Cush JJ, Mejias A, Ramilo O, Kon OM, Pascual V, Banchereau J, Chaussabel D, O'Garra A (2010) An interferon-inducible neutrophil-driven blood transcriptional signature in human tuberculosis. Nature 466: 973–977
46. Kirou KA, Lee C, George S, Louca K, Peterson MG, Crow MK (2005) Activation of the interferon-alpha pathway identifies a subgroup of systemic lupus erythematosus patients with distinct serologic features and active disease. Arthritis Rheum 52:1491–1503
47. Baechler EC, Batliwalla FM, Karypis G, Gaffney PM, Ortmann WA, Espe KJ, Shark KB, Grande WJ, Hughes KM, Kapur V, Gregersen PK, Behrens TW (2003) Interferon-inducible gene expression signature in peripheral blood cells of patients with severe lupus. Proc Natl Acad Sci U S A 100:2610–2615
48. Bennett L, Palucka AK, Arce E, Cantrell V, Borvak J, Banchereau J, Pascual V (2003) Interferon and granulopoiesis signatures in systemic lupus erythematosus blood. J Exp Med 197:711–723
49. Ottenhoff TH, Dass RH, Yang N, Zhang MM, Wong HE, Sahiratmadja E, Khor CC, Alisjahbana B, van Crevel R, Marzuki S, Seielstad M, van de Vosse E, Hibberd ML (2012) Genome-wide expression profiling identifies type 1 interferon response pathways in active tuberculosis. PLoS One 7:e45839
50. Wu K, Dong D, Fang H, Levillain F, Jin W, Mei J, Gicquel B, Du Y, Wang K, Gao Q, Neyrolles O, Zhang J (2012) An interferon-related signature in the transcriptional core response of human macrophages to *Mycobacterium tuberculosis* infection. PLoS One 7:e38367
51. Maertzdorf J, Weiner J 3rd, Mollenkopf HJ, Network TB, Bauer T, Prasse A, Muller-Quernheim J, Kaufmann SH (2012) Common patterns and disease-related signatures in tuberculosis and sarcoidosis. Proc Natl Acad Sci U S A 109:7853–7858
52. Koh GC, Schreiber MF, Bautista R, Maude RR, Dunachie S, Limmathurotsakul D, Day NP, Dougan G, Peacock SJ (2013) Host responses to melioidosis and tuberculosis are both dominated by interferon-mediated signaling. PLoS One 8:e54961

53. Wang J, Zhou X, Pan B, Wang H, Shi F, Gan W, Yang L, Yin X, Xu B, Zhao D (2013) Expression pattern of interferon-inducible transcriptional genes in neutrophils during bovine tuberculosis infection. DNA Cell Biol 32:480–486
54. Berry MP, Blankley S, Graham CM, Bloom CI, O'Garra A (2013) Systems approaches to studying the immune response in tuberculosis. Curr Opin Immunol 25:579–587
55. Bloom CI, Graham CM, Berry MP, Wilkinson KA, Oni T, Rozakeas F, Xu Z, Rossello-Urgell J, Chaussabel D, Banchereau J, Pascual V, Lipman M, Wilkinson RJ, O'Garra A (2012) Detectable changes in the blood transcriptome are present after two weeks of antituberculosis therapy. PLoS One 7:e46191
56. Ordway D, Henao-Tamayo M, Harton M, Palanisamy G, Troudt J, Shanley C, Basaraba RJ, Orme IM (2007) The hypervirulent *Mycobacterium tuberculosis* strain HN878 induces a potent TH1 response followed by rapid down-regulation. J Immunol 179:522–531
57. Desvignes L, Wolf AJ, Ernst JD (2012) Dynamic roles of type I and type II IFNs in early infection with *Mycobacterium tuberculosis*. J Immunol 188:6205–6215
58. Antonelli LR, Gigliotti Rothfuchs A, Goncalves R, Roffe E, Cheever AW, Bafica A, Salazar AM, Feng CG, Sher A (2010) Intranasal poly-IC treatment exacerbates tuberculosis in mice through the pulmonary recruitment of a pathogen-permissive monocyte/macrophage population. J Clin Invest 120:1674–1682
59. Redford PS, Mayer-Barber KD, McNab FW, Stavropoulos E, Wack A, Sher A, O'Garra A (2014) Influenza A virus impairs control of *Mycobacterium tuberculosis* coinfection through a type I interferon receptor-dependent pathway. J Infect Dis 209:270–274
60. Bengtsson AA, Sturfelt G, Truedsson L, Blomberg J, Alm G, Vallin H, Ronnblom L (2000) Activation of type I interferon system in systemic lupus erythematosus correlates with disease activity but not with antiretroviral antibodies. Lupus 9:664–671
61. Dall'era MC, Cardarelli PM, Preston BT, Witte A, Davis JC Jr (2005) Type I interferon correlates with serological and clinical manifestations of SLE. Ann Rheum Dis 64: 1692–1697
62. Erdozain JG, Ruiz-Irastorza G, Egurbide MV, Martinez-Berriotxoa A, Aguirre C (2006) High risk of tuberculosis in systemic lupus erythematosus? Lupus 15:232–235
63. Prabu V, Agrawal S (2010) Systemic lupus erythematosus and tuberculosis: a review of complex interactions of complicated diseases. J Postgrad Med 56:244–250
64. Sayarlioglu M, Inanc M, Kamali S, Cefle A, Karaman O, Gul A, Ocal L, Aral O, Konice M (2004) Tuberculosis in Turkish patients with systemic lupus erythematosus: increased frequency of extrapulmonary localization. Lupus 13:274–278
65. Mitchell SR, Nguyen PQ, Katz P (1990) Increased risk of neisserial infections in systemic lupus erythematosus. Semin Arthritis Rheum 20:174–184
66. Zandman-Goddard G, Shoenfeld Y (2003) SLE and infections. Clin Rev Allergy Immunol 25:29–40
67. Yao Y, Richman L, Higgs BW, Morehouse CA, de los Reyes M, Brohawn P, Zhang J, White B, Coyle AJ, Kiener PA, Jallal B (2009) Neutralization of interferon-alpha/beta-inducible genes and downstream effect in a phase I trial of an anti-interferon-alpha monoclonal antibody in systemic lupus erythematosus. Arthritis Rheum 60:1785–1796
68. Pestka S, Krause CD, Walter MR (2004) Interferons, interferon-like cytokines, and their receptors. Immunol Rev 202:8–32
69. Sabbatani S, Manfredi R, Marinacci G, Pavoni M, Cristoni L, Chiodo F (2006) Reactivation of severe, acute pulmonary tuberculosis during treatment with pegylated interferon-alpha and ribavirin for chronic HCV hepatitis. Scand J Infect Dis 38:205–208
70. Belkahla N, Kchir H, Maamouri N, Ouerghi H, Hariz FB, Chouaib S, Chaabouni H, Mami NB (2010) [Reactivation of tuberculosis during dual therapy with pegylated interferon and ribavirin for chronic hepatitis C]. Rev Med Interne 31:e1–3
71. Farah R, Awad J (2007) The association of interferon with the development of pulmonary tuberculosis. Int J Clin Pharmacol Ther 45:598–600

72. Telesca C, Angelico M, Piccolo P, Nosotti L, Morrone A, Longhi C, Carbone M, Baiocchi L (2007) Interferon-alpha treatment of hepatitis D induces tuberculosis exacerbation in an immigrant. J Infect 54:e223–e226
73. Perez-Elias MJ, Garcia-San Miguel L, Gonzalez Garcia J, Montes Ramirez ML, Muriel A, Machin-Lazaro JM, Martinez-Baltanas A, Zamora F, Moreno A, Martin-Davila P, Quereda C, Gomez-Mampaso E, Moreno S (2009) Tuberculosis complicating hepatitis C treatment in HIV-infected patients. Clin Infect Dis 48:e82–e85
74. Tsai MC, Lin MC, Hung CH (2009) Successful antiviral and antituberculosis treatment with pegylated interferon-alfa and ribavirin in a chronic hepatitis C patient with pulmonary tuberculosis. J Formos Med Assoc 108:746–750
75. Giosue S, Casarini M, Alemanno L, Galluccio G, Mattia P, Pedicelli G, Rebek L, Bisetti A, Ameglio F (1998) Effects of aerosolized interferon-alpha in patients with pulmonary tuberculosis. Am J Respir Crit Care Med 158:1156–1162
76. Zarogoulidis P, Kioumis I, Papanas N, Manika K, Kontakiotis T, Papagianis A, Zarogoulidis K (2012) The effect of combination IFN-alpha-2a with usual antituberculosis chemotherapy in non-responding tuberculosis and diabetes mellitus: a case report and review of the literature. J Chemother 24:173–177
77. Pecora ND, Gehring AJ, Canaday DH, Boom WH, Harding CV (2006) *Mycobacterium tuberculosis* LprA is a lipoprotein agonist of TLR2 that regulates innate immunity and APC function. J Immunol 177:422–429
78. Akira S, Takeda K (2004) Toll-like receptor signalling. Nat Rev Immunol 4:499–511
79. Yamamoto M, Sato S, Hemmi H, Hoshino K, Kaisho T, Sanjo H, Takeuchi O, Sugiyama M, Okabe M, Takeda K, Akira S (2003) Role of adaptor TRIF in the MyD88-independent toll-like receptor signaling pathway. Science 301:640–643
80. Simmons DP, Canaday DH, Liu Y, Li Q, Huang A, Boom WH, Harding CV (2010) *Mycobacterium tuberculosis* and TLR2 agonists inhibit induction of type I IFN and class I MHC antigen cross processing by TLR9. J Immunol 185:2405–2415
81. Park JH, Kim YG, McDonald C, Kanneganti TD, Hasegawa M, Body-Malapel M, Inohara N, Nunez G (2007) RICK/RIP2 mediates innate immune responses induced through Nod1 and Nod2 but not TLRs. J Immunol 178:2380–2386
82. Monroe KM, McWhirter SM, Vance RE (2010) Induction of type I interferons by bacteria. Cell Microbiol 12:881–890
83. Watanabe T, Asano N, Fichtner-Feigl S, Gorelick PL, Tsuji Y, Matsumoto Y, Chiba T, Fuss IJ, Kitani A, Strober W (2010) NOD1 contributes to mouse host defense against *Helicobacter pylori* via induction of type I IFN and activation of the ISGF3 signaling pathway. J Clin Invest 120:1645–1662
84. Leber JH, Crimmins GT, Raghavan S, Meyer-Morse NP, Cox JS, Portnoy DA (2008) Distinct TLR- and NLR-mediated transcriptional responses to an intracellular pathogen. PLoS Pathog 4:e6
85. Divangahi M, Mostowy S, Coulombe F, Kozak R, Guillot L, Veyrier F, Kobayashi KS, Flavell RA, Gros P, Behr MA (2008) NOD2-deficient mice have impaired resistance to *Mycobacterium tuberculosis* infection through defective innate and adaptive immunity. J Immunol 181: 7157–7165
86. Watson RO, Manzanillo PS, Cox JS (2012) Extracellular *M. tuberculosis* DNA targets bacteria for autophagy by activating the host DNA-sensing pathway. Cell 150:803–815
87. Ishikawa H, Ma Z, Barber GN (2009) STING regulates intracellular DNA-mediated, type I interferon-dependent innate immunity. Nature 461:788–792
88. Ishikawa H, Barber GN (2011) The STING pathway and regulation of innate immune signaling in response to DNA pathogens. Cell Mol Life Sci 68:1157–1165
89. Barber GN (2011) Innate immune DNA sensing pathways: STING, AIMII and the regulation of interferon production and inflammatory responses. Curr Opin Immunol 23:10–20
90. Karaolis DK, Means TK, Yang D, Takahashi M, Yoshimura T, Muraille E, Philpott D, Schroeder JT, Hyodo M, Hayakawa Y, Talbot BG, Brouillette E, Malouin F (2007) Bacterial c-di-GMP is an immunostimulatory molecule. J Immunol 178:2171–2181

91. McWhirter SM, Barbalat R, Monroe KM, Fontana MF, Hyodo M, Joncker NT, Ishii KJ, Akira S, Colonna M, Chen ZJ, Fitzgerald KA, Hayakawa Y, Vance RE (2009) A host type I interferon response is induced by cytosolic sensing of the bacterial second messenger cyclic-di-GMP. J Exp Med 206:1899–1911
92. Woodward JJ, Iavarone AT, Portnoy DA (2010) c-di-AMP secreted by intracellular *Listeria monocytogenes* activates a host type I interferon response. Science 328:1703–1705
93. Burdette DL, Monroe KM, Sotelo-Troha K, Iwig JS, Eckert B, Hyodo M, Hayakawa Y, Vance RE (2011) STING is a direct innate immune sensor of cyclic di-GMP. Nature 478: 515–518
94. Stewart GR, Patel J, Robertson BD, Rae A, Young DB (2005) Mycobacterial mutants with defective control of phagosomal acidification. PLoS Pathog 1:269–278
95. Cui T, He Z (2012) C-di-GMP signaling and implications for pathogenesis of *Mycobacterium tuberculosis*. Chin Sci Bull 57:4387–4393
96. Mayer-Barber KD, Andrade BB, Barber DL, Hieny S, Feng CG, Caspar P, Oland S, Gordon S, Sher A (2011) Innate and adaptive interferons suppress IL-1alpha and IL-1beta production by distinct pulmonary myeloid subsets during *Mycobacterium tuberculosis* infection. Immunity 35:1023–1034
97. Beamer GL, Flaherty DK, Assogba BD, Stromberg P, Gonzalez-Juarrero M, de Waal Malefyt R, Vesosky B, Turner J (2008) Interleukin-10 promotes *Mycobacterium tuberculosis* disease progression in CBA/J mice. J Immunol 181:5545–5550
98. Redford PS, Boonstra A, Read S, Pitt J, Graham C, Stavropoulos E, Bancroft GJ, O'Garra A (2010) Enhanced protection to *Mycobacterium tuberculosis* infection in IL-10-deficient mice is accompanied by early and enhanced Th1 responses in the lung. Eur J Immunol 40: 2200–2210
99. Redford PS, Murray PJ, O'Garra A (2011) The role of IL-10 in immune regulation during *M. tuberculosis* infection. Mucosal Immunol 4:261–270
100. Piganis RA, De Weerd NA, Gould JA, Schindler CW, Mansell A, Nicholson SE, Hertzog PJ (2011) Suppressor of cytokine signaling (SOCS) 1 inhibits type I interferon (IFN) signaling via the interferon alpha receptor (IFNAR1)-associated tyrosine kinase Tyk2. J Biol Chem 286:33811–33818
101. Fenner JE, Starr R, Cornish AL, Zhang JG, Metcalf D, Schreiber RD, Sheehan K, Hilton DJ, Alexander WS, Hertzog PJ (2006) Suppressor of cytokine signaling 1 regulates the immune response to infection by a unique inhibition of type I interferon activity. Nat Immunol 7: 33 39
102. Almeida AS, Lago PM, Boechat N, Huard RC, Lazzarini LC, Santos AR, Nociari M, Zhu H, Perez-Sweeney BM, Bang H, Ni Q, Huang J, Gibson AL, Flores VC, Pecanha LR, Kritski AL, Lapa e Silva JR, Ho JL (2009) Tuberculosis is associated with a down-modulatory lung immune response that impairs Th1-type immunity. J Immunol 183:718–731
103. Masood KI, Rottenberg ME, Salahuddin N, Irfan M, Rao N, Carow B, Islam M, Hussain R, Hasan Z (2013) Expression of *M. tuberculosis*-induced suppressor of cytokine signaling (SOCS) 1, SOCS3, FoxP3 and secretion of IL-6 associates with differing clinical severity of tuberculosis. BMC Infect Dis 13:13
104. Carow B, Ye X, Gavier-Widen D, Bhuju S, Oehlmann W, Singh M, Skold M, Ignatowicz L, Yoshimura A, Wigzell H, Rottenberg ME (2011) Silencing suppressor of cytokine signaling-1 (SOCS1) in macrophages improves *Mycobacterium tuberculosis* control in an interferon-gamma (IFN-gamma)-dependent manner. J Biol Chem 286:26873–26887
105. Kearney SJ, Delgado C, Eshleman EM, Hill KK, O'Connor BP, Lenz LL (2013) Type I IFNs downregulate myeloid cell IFN-gamma receptor by inducing recruitment of an early growth response 3/NGFI-A binding protein 1 complex that silences Ifngr1 transcription. J Immunol 191:3384–3392
106. Rayamajhi M, Humann J, Penheiter K, Andreasen K, Lenz LL (2010) Induction of IFN-alphabeta enables *Listeria monocytogenes* to suppress macrophage activation by IFN-gamma. J Exp Med 207:327–337

107. Bozza PT, Bai X, Feldman NE, Chmura K, Ovrutsky AR, Su W-L, Griffin L, Pyeon D, McGibney MT, Strand MJ, Numata M, Murakami S, Gaido L, Honda JR, Kinney WH, Oberley-Deegan RE, Voelker DR, Ordway DJ, Chan ED (2013) Inhibition of nuclear factor-kappa B activation decreases survival of *Mycobacterium tuberculosis* in human macrophages. PLoS One 8:e61925
108. McNab FW, Ewbank J, Rajsbaum R, Stavropoulos E, Martirosyan A, Redford PS, Wu X, Graham CM, Saraiva M, Tsichlis P, Chaussabel D, Ley SC, O'Garra A (2013) TPL-2-ERK1/2 signaling promotes host resistance against intracellular bacterial infection by negative regulation of type I IFN production. J Immunol 191:1732–1743

Index

© Springer International Publishing Switzerland 2014
D. Parker (ed.), *Bacterial Activation of Type I Interferons*,
DOI 10.1007/978-3-319-09498-4